Bioremediation for Global Environmental Conservation

Edited by Naofumi Shiomi, Vasudeo Zambare and Mohd Fadhil Md. Din

Published in London, United Kingdom

Bioremediation for Global Environmental Conservation
http://dx.doi.org/10.5772/intechopen.107789
Edited by Naofumi Shiomi, Vasudeo Zambare and Mohd Fadhil Md. Din

Contributors

Radhika Birmole, Aruna K. Samudravijay, Samadhan R. Waghmode, Anup Sonawane, Elva Teresa Arechiga-Carvajal, Juan Cabral-Miramontes, Pamela Dorantes-Alvarado, Maria Banda, Wilma Augustyn, Alexis Munyengabe, Vladimir Arias-Arce, Daniel Lovera-Dávila, José J. Guerrero-Rojas, Fanny Blas-Rodríguez, Ismael Molina-Pereyra, Ioana Stanciu, Suzan P. Vasconcellos, André Paganotti, Vitor G. Vital, Lidiane M. Santos Lima, Giovanna S.M. Paiva, L. Furlaneto De Lima, Enrique Moreira, Leticia O. Sousa, Guilherme G. Guerini, Vinicius T. Santos, Flavia G. Lobo, Márcio R. Silva, Diogo S. Pellosi, Ricardo Alexandre A.G. Silva, Vasudeo Zambare, Mohd Fadhil Md. Din, Naofumi Shiomi

Notice

Statements and opinions expressed in the chapters are these of the individual contributors and not necessarily those of the editors or publisher. No responsibility is accepted for the accuracy of information contained in the published chapters. The publisher assumes no responsibility for any damage or injury to persons or property arising out of the use of any materials, instructions, methods or ideas contained in the book.

First published in London, United Kingdom, 2023 by IntechOpen
IntechOpen is the global imprint of INTECHOPEN LIMITED, registered in England and Wales, registration number: 11086078, 5 Princes Gate Court, London, SW7 2QJ, United Kingdom

British Library Cataloguing-in-Publication Data
A catalogue record for this book is available from the British Library

Additional hard and PDF copies can be obtained from orders@intechopen.com

Bioremediation for Global Environmental Conservation
Edited by Naofumi Shiomi, Vasudeo Zambare and Mohd Fadhil Md. Din
p. cm.

This title is part of the Environmental Sciences Book Series, Volume 12
Topic: Environmental Resilience and Management
Series Editor:J. Kevin Summers
Topic Editor: Jose Navarro-Pedreño

Print ISBN 978-1-83768-981-1
Online ISBN 978-1-83768-982-8
eBook (PDF) ISBN 978-1-83768-983-5
ISSN 2754-6713

For EU product safety concerns:
IN TECH d.o.o., Prolaz Marije Krucifikse Kozulić 3, 51000 Rijeka, Croatia,
info@intechopen.com or visit our website at intechopen.com.

IntechOpen Book Series
Environmental Sciences
Volume 12

Aims and Scope of the Series

Scientists have long researched to understand the environment and man's place in it. The search for this knowledge grows in importance as rapid increases in population and economic development intensify humans' stresses on ecosystems. Fortunately, rapid increases in multiple scientific areas are advancing our understanding of environmental sciences. Breakthroughs in computing, molecular biology, ecology, and sustainability science are enhancing our ability to utilize environmental sciences to address real-world problems.

The four topics of this book series - Pollution; Environmental Resilience and Management; Ecosystems and Biodiversity; and Water Science - will address important areas of advancement in the environmental sciences. They will represent an excellent initial grouping of published works on these critical topics.

Meet the Series Editor

J. Kevin Summers is a Senior Research Ecologist at the Environmental Protection Agency's (EPA) Gulf Ecosystem Measurement and Modeling Division. He is currently working with colleagues in the Sustainable and Healthy Communities Program to develop an index of community resilience to natural hazards, an index of human well-being that can be linked to changes in the ecosystem, social and economic services, and a community sustainability tool for communities with populations under 40,000. He leads research efforts for indicator and indices development. Dr. Summers is a systems ecologist and began his career at the EPA in 1989 and has worked in various programs and capacities. This includes leading the National Coastal Assessment in collaboration with the Office of Water which culminated in the award-winning National Coastal Condition Report series (four volumes between 2001 and 2012), and which integrates water quality, sediment quality, habitat, and biological data to assess the ecosystem condition of the United States estuaries. He was acting National Program Director for Ecology for the EPA between 2004 and 2006. He has authored approximately 150 peer-reviewed journal articles, book chapters, and reports and has received many awards for technical accomplishments from the EPA and from outside of the agency. Dr. Summers holds a BA in Zoology and Psychology, an MA in Ecology, and Ph.D. in Systems Ecology/Biology.

Meet the Volume Editors

Dr. Naofumi Shiomi studied recombinant yeast as a researcher at the Laboratory of Production Technology of Kaneka Corporation for 15 years until 1998 and earned his Ph.D. in Engineering from Kyoto University, Japan. He now works as a professor at the School of Human Sciences, Kobe College, Japan, where he teaches applied microbiology, biotechnology, and life science, and studies at his Applied Science laboratories. He has studied bioremediation for 20 years at Kobe College. His research during the last decade has focused on cellular rejuvenation and prevention of metabolic syndrome.

Dr. Vasudeo Zambare is a Universiti Teknologi Malaysia (UTM) Fellow at the Center for Environmental Sustainability and Water Security (IPASA), UTM, and the head of R&D at Balaji Enzyme and Chemical Pvt Ltd, Mumbai, Maharashtra, India. He obtained his Ph.D. in Biochemistry from Agharkar Research Institute, Pune University, Maharashtra, India, in 2007. He has 3 patents and more than 230 technical and scientific contributions and 80 peer-reviewed journal articles to his credit. Dr. Zambare is a multi-skilled researcher with biorefinery industry experience in the United States, Canada, the European Union, India, and Malaysia. He has developed bioprocesses for the leather, textile, paper, and pulp, and biofuel industries. He has more than 20 years of achievements in advancing knowledge by devising fermentation process development, assays, and analytical methods to solve complex research problems with potential commercial applications (biofuel, food, and pharmaceuticals). Dr. Zambare's research expertise includes industrial enzymes, probiotics, extremophiles, biofertilizers, biopesticides, waste management, and leather bioprocessing.

Dr. Mohd Fadhil Md. Din graduated from Universiti Teknologi Malaysia (UTM) in 1999, also holding a Master of Engineering in 2001. He further studied under a double program between UTM and Delft University of Technology (TUDelft), Netherlands. Dr. Din has more than 280 technical and scientific contributions to his credit, including book chapters, books, proceedings, popular articles, conferences, and workshops. He has received several funding grants and prestigious awards. Dr. Din is a director of campus sustainability and works for environmental sustainability in the campus. He has expertise in environmental sciences and applications, environmental management systems (EMS), environmental impact assessment (EIA), risk management, and transportation projects (mobility transformational agenda). He develops academic learning curriculum and training courses in various types of environmental science and engineering subjects for undergraduate and postgraduate courses, specializing in water and wastewater treatment, biotechnology and bioengineering, fundamental chemical sciences, and river rehabilitation, particularly in developing countries. He has received several awards and recognitions at both national and international levels.

Contents

Preface

The concentration of carbon dioxide (CO_2) in the atmosphere has continued to increase rapidly in the last hundred years, and the global average temperature is predicted to rise by 1.5°C by 2027 and by 3.3–5.7°C by the end of this century. If global average temperatures rise by 5°C or more, many species will not be able to live, as there will be extremely high temperatures in the summer as well as natural disasters such as storms, floods, and water shortages. Submergence will also occur with unusual frequency because of the rise in seawater temperature. These environmental changes would not only cause a fatal blow to the world economy but also would likely render some parts of the earth uninhabitable.

Furthermore, global warming has also weakened the Atlantic meridional overturning circulation (AMOC), an ocean system that flows in the north-south direction of the Atlantic Ocean. A team from the University of Copenhagen in Denmark predicts that the AMOC will likely cease by as early as 2025 or as late as 2095. If the AMOC stops, humans will experience extreme weather events, such as tropical rainforests moving south and Europe drying out. Water shortages due to climate change as well as soil and groundwater contamination due to environmental pollution are serious issues, with reports in 2007 indicating that people in about 30 to 40 countries (660 million people) are suffering from extreme water shortages; and the number is rapidly increasing. In addition, soil contamination is causing a rapid decrease in farmland available for food production, which is likely to be further exacerbated if natural disasters caused by global warming intensify, resulting in severe food shortages in the near future. It is the urgent task of the world's scientists to solve these problems by the end of the century through the mobilization of science and technology to prevent global warming by reducing CO_2 emissions, developing alternative energy sources to fossil fuels, and securing food and drinking water by remediating soil and groundwater contamination. Biotechnology has a major role to play regarding the same.

Against this background, this book focuses on the remediation of widespread pollution, which is one of the most important issues for maintaining a sustainable society. The views of many experts on new strategies for efficient remediation and the material in this book provide important information for the rapid cleanup of contaminated soil and groundwater.

The authors would like to thank Mr. Dominik Samardzija, Ms. Iva Simcic Mancic, and the publishing staff at IntechOpen for their invaluable assistance in writing and publishing this book.

Naofumi Shiomi
School of Human Sciences,
Kobe College,
Nishinomiya, Japan

Vasudeo Zambare
Center for Environmental Sustainability and Water Security (IPASA),
Universiti Teknologi Malaysia,
Johor Bahru, Malasia

R&D Department,
Balaji Enzyme and Chemical Pvt Ltd,
Mumbai, India

Mohd Fadhil Md. Din
Center for Environmental Sustainability and Water Security (IPASA)
and Faculty of Civil Engineering,
Universiti Teknologi Malaysia,
Johor Bahru, Malasia

Chapter 1

Introductory Chapter: Biodegradation – New Insights

Vasudeo Zambare and Mohd Fadhil Md. Din

1. Introduction

Microbial breakdown of any matter or compound is called biodegradation, which is carried out by bacteria, fungi, or yeasts [1]. Though it is a natural process but it is somewhat different than a composting process. Composting is a human-interfered systematic and safe process where organic matter is a breakdown within a specific time, whereas biodegradability is the conversion where it is not necessary that it will be beneficial always [2]. All natural organic matters and chemical compounds are biodegradable provided with time as a major constraint. Organic materials, including food wastes and vegetables, are degraded in few days; however, the plastic degradation may take several years.

Biodegradation occurs in three stages of biodeterioration, biofragmentation, and bioassimilation [3]. Biodeterioration is the superficial but surface-level degradation *via* alteration of all physical, chemical, and mechanical properties of the material. This process is generally occurred abiotically where environmental factors, including light, temperature, chemical exposure, and mechanical compressions, weakening or deteriorate the material. Thus, biodeterioration is the first stage of biodegradation and is in parallel to the microbial action for the fragmentation of the material. Biofragmentation is a very crucial and important stage where high-molecular-weight molecules or polymeric structures are disturbed by either bond cleavage by enzymatic action or chemical modification. Resultant into a generation of oligomeric or monomeric fragments. This biofragmentation is an oxidation-reduction process and energy intensive. The biofragmentation process that occurs in the presence of oxygen is called aerobic degradation, and if it occurs in the absence of oxygen, then it is called anaerobic degradation. Comparatively, each process generates water and CO_2; however, methane and hydrogen are strictly produced by anaerobic process only. Aerobic digestion is a rapid process, but anaerobic digestion is more efficient in terms of mass reduction by producing natural gas. Hence, the anaerobic process is most widely used in waste management and for renewable energy generation. Bioassimilation is the last stage of biodegradation where microbes utilize the bio-fragmented material for their metabolic activities either for energy generation (adenosine triphosphate, ATP) or as precursors of any other biosynthetic pathways.

Factors affecting material biodegradation are water, light, oxygen, and temperature. For biodegradation of any organic compound, it is necessary to make that compound available for microbial action where the compound is absorbed by the system. Biodegradability of solid and liquid is also differing where CO_2 and methane gas are produced during aerobic and anaerobic digestion processes, respectively. Factorial studies in the lab must have their working feasibility in the field and hence sometimes

 IntechOpen

lab results are not workable in a large scale. Considering the biodegradation of landfilled solid waste, where physical factors of water, light, and microbial activities are varies and are directly affecting the biodegradability efficiencies. And plastic degradation which is environmentally dependent is altogether very different and completely unpredictable. It is also important to find out the methods of degradability and one such method is DINV 54900 [4]. Bioremediation is another form of biodegradability where toxic chemicals are converted into simpler form and the best examples are the textile dyes. Our earlier studies showed decolorization of textile dyes using microbial cultures and contributed to environmental protection by remediating the textile dyed effluent [5, 6].

2. Biodegradability

Biodegradability of any compound is measured by the respirometer method which measured the gas (CO_2), which is a metabolic product from microbes, especially

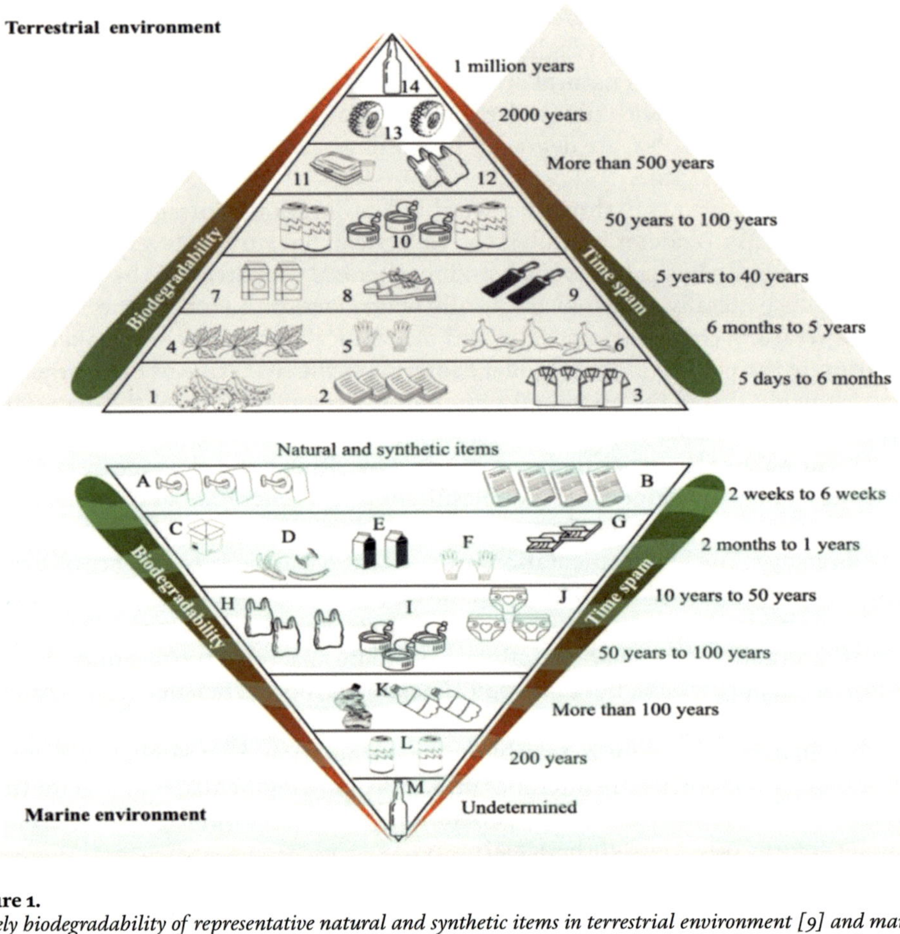

Figure 1.
Timely biodegradability of representative natural and synthetic items in terrestrial environment [9] and marine environment [10] (1, food and vegetables; 2, cellulose and paper; 3, cotton cloths; 4, plant leaves; 5, wool products; 6, fruit peels; 7, plastic coated food boxes; leather products; 9, nylon products; 10, tin and aluminum cans; 11, Styrofoam products; 12, plastic bags and items; 13, vehicle tires; 14, glass items; A, tissue paper (toilet); B, daily newspaper; C, cardboard; D, fruit wastage; E, wax coated milk carton; F, cotton & woolen gloves; G, plywood; H, plastic bags; I, tin cans; J, diapers; K, plastic bottles; L, aluminum cans; M, glass items).

aerobic microbes [7]. The present article is discussing the time-wise trend of biodegradability of some representative natural and synthetic items under terrestrial and marine environment. Biodegradability is depending on light, water, temperature, oxygen, and of course microbes, which is widely distributed and differentiated on the terrestrial and marine environment [8]. Hence, the degradability period would be differently affected. **Figure 1** is showing an integration of biodegradability trends in both terrestrial and marine environment and looks like marine environment showed more efficient biodegradability. That might be due to a lot of shear and stress conditions as well as climatic conditions of availability of light, oxygen tides frequencies, and diversified microflora. Though terrestrial environment has some limitations of light availability in case of heap or land filing but each environment has its advantages and limitations. Biodegradation occurs through various processes such as biotransformation of toxic to nontoxic compounds [11], volume reduction *via* biodeterioration by microbial and enzymatic action, formation of minerals *via* biomineralization [12], convert waste to another usable or high-value form *via* recycling or valorization [13]. Thus, all forms of material will have biodegradability but it is time-dependent. Some might take few weeks to several years. Organic materials like plant-based fruits, vegetables, and lignocellulosic materials can be easily degraded and are based on their structure and crystallinity [14]. There are some semi-synthetic and metallic materials that took almost 100 years for degradation; however, the toughest and harder to degrade in both terrestrial and marine environment are plastic, rubber, and glass materials [15]. As long as rubber is a natural material, it is biodegradable but once it is converted into a crosslinked tire (as composite), then it takes more than 2000 years to degrade it [16]. Glass is a natural and very toughest material where no biodegradability is predicted. But the most alarming and dangerous is the plastic which is 100% synthetic.

3. Challenges and opportunities

High biodegradability of products is always prioritized but products with no or low biodegradability are interfering in agriculture, human, animal health, and aquatic life. Especially, plastic biodegradability is a major concern. Domestic and wildlife animals accidentally eat the plastic as food items and showed entanglement in the intestine [17]. Polyhalogenated compounds, pesticides, dyes, hydrocarbons are slowly degrading and also interfere in the animal food chain. Indirectly, these chemicals showed very slow but bad impacts on human health by biomagnification and bioaccumulation process of food which ended in serious health complications such as cancer, neurological disorder, and hormonal dysfunction [18]. Mercury is one of the known contaminants and some fish with high mercury affected the human hormonal problem when entered in the human food chain [19]. Overall, nonbiodegradable items are real challenges for researchers and scientists. Many researchers have initiated biodegradation research on plastic [20–23]. These findings showed various bacterial and fungal strains for plastic biodegradation by enzymatic as well as photodegradation and thermo-oxidative activities. Polyethylene (PET) is one of the widely spread and a major contaminating plastic present everywhere but with research efforts, its degradation has been identified by *Ideonella sakaiensis* with an enzyme PETase [24]. With the advancement and use of genetic engineering, the PETase has been modified with MHETase for fast degradation of PET as well as polyethylene furan-2,5 dicarxylate (PEF) [24]. Adaptability of microbial strains might be altering their metabolic pathways and thus resulting into the findings of plastic degrading enzyme secretions which is also evidenced by Quartinello et al. [25] from cow gut microbes for breakdown of three different types of plastic. Though research is in the direction of positive output but more efforts and multilevel research methodologies need to be utilized for innovation-based findings related to novel strains, biocatalyst, or genetic modification. One more challenge of formation of microplastic and nanoplastic which is ended in the marine environment and acts as major contaminants affecting the marine flora [26].

Another opportunity for development of biodegradable alternatives but the major limitations are the enough tensile and mechanical strength. With a wider scope, cellulose, starch, chitin, based materials, and some polymers such as polyhydroxyalkanoates, polyhydroxybutyrate, polylactic acid, and polycaprolactone are some of the highly biodegradable polymeric materials obtained from plants and microbes [27]. Again, these microbial and plant-based polymers have some limitations in stability, mouldability, and consistency. Hence, an opportunity for a composite technology for novel material of desirable properties as well as biodegradability is open. A technology related to biodegradability of material needs to be developed for packaging, production, and medicine using some oxo-biodegradable formulations. Sometimes, the biodegradability of material generates greenhouse gases and has an opportunity to valorize this polymer to another but more useful and nontoxic precursor. Overall, there are several opportunities to develop cutting edge research on existing plastic like tough material degradation or investigate a new microbial strain or modification in existing strain for more efficient and 100% biodegradable materials having affordable economic and industrial feasibilities.

4. Conclusions

Biodegradability "deals with the waste management concept where proper disposal and reduction in volume" is required with the use of completely biodegradable products or the need to develop a robust biocatalyst from a microbial system.

Acknowledgements

Authors acknowledge special thanks to the Research Management Center of Universiti Teknologi Malaysia, Malaysia for Fellow Research Grant award (PY/2022/03188).

Conflict of interest

All authors have no conflict of interest.

Author details

Vasudeo Zambare[1,2*] and Mohd Fadhil Md. Din[2,3]

1 R&D Department, Om Biotechnologies, Nashik, Maharashtra, India

2 Centre for Environmental Sustainability and Water Security (IPASA), Universiti Teknologi Malaysia, Johor Bahru, Malaysia

3 Department of Water and Environmental Engineering, Universiti Teknologi Malaysia, Johor Bahru, Malaysia

*Address all correspondence to: vasuzambare@gmail.com; vpzambare@utm.my

References

[1] Tahri N, Bahafid W, Sayel H, et al. Biodegradation: Involved microorganisms and genetically engineered microorganisms. Biodegradation Life of Science. 2013. DOI: 10.5772/56194

[2] Mouhoubi R, Lasschuijt M, Ramon CS, et al. End-of-life biodegradation? How to assess the composting of polyesters in the lab and the field. Waste Management. 2022;**154**:36-48. DOI: 10.1016/j.wasman.2022.09.025

[3] Glaser AJ. Biological degradation of polymers in the environment. IntechOpen. 2019. DOI: 10.5772/intechopen.85124

[4] Witt U, Yamamoto M, Seeliger U, et al. Biodegradable polymeric materials-not the origin but the chemical structure determines biodegradability. Angewandte Chemie. 1999;**38**(10):1438-1442. DOI: 10.1002/(sici)1521-3773(19990517)38

[5] Ponraj M, Gokila K, Zambare VP. Bacterial decolorization of textile dye-Orange 3R. International Journal of Advanced Biotechnology and Research. 2011;**2**(1):168-177

[6] Ponraj M, Jamunarani P, Zambare V. Isolation and optimization of culture conditions for decolorization of true-blue using dye decolorizing fungi. Asian Journal of Experimental and Biological Sciences. 2011;**2**(2):270-277

[7] Aichinger G, Grady CPL, Tabak HH. Application of respirometric biodegradability testing protocol to slightly soluble organic compounds. Water Environment Research. 1992;**64**(7):890-900. Available from: http://www.jstor.org/stable/25044246

[8] Chamas A, Moon H, Zheng J, et al. Degradation rates of plastics in the environment. ACS Sustainable Chemical Engineering. 2020;**8**(9):3494-3511. DOI: 10.1021/acssuschemeng.9b06635

[9] Science Learning Hub. Measuring Biodegradability. Hamilton, New Zealand: Science Learning Hub; 2008. Available from: https://www.sciencelearn.org.nz/resources/1543-measuring-biodegradability [Accessed: 2 April, 2023]

[10] More C. Marine debris biodegradation timeline. Mote Marine Laboratory. 1993. Available from: http://cmore.soest.hawaii.edu/cruises/super/biodegradation.html [Accessed: 2 April, 2023]

[11] De Sousa IP, Teixeira MVS, Furtado NAJC. An overview of biotransformation and toxicity of diterpenes. Molecules. 2018;**23**(6):1387. DOI: 10.3390/molecules23061387

[12] Ben-Tabou de-Leon S. The evolution of biomineralization through the co-option of organic scaffold forming networks. Cells. 2018;**11**(4):595. DOI: 10.3390/cells11040595

[13] Pereira F, Silva C. Energetic valorization of bio-waste from municipal solid waste in Porto Santo Island. Clean Technologies. 2023;**5**(1):233-258. DOI: 10.3390/cleantechnol5010014

[14] Suryadi H, Judono JJ, Putri MR, et al. Biodelignification of lignocellulose using ligninolytic enzymes from white-rot fungi. Heliyon. 2022;**8**(2):e08865. DOI: 10.1016/j.heliyon.2022.e08865

[15] Thushari GGN, Senevirathna JDM. Plastic pollution in the marine environment. Heliyon. 2020;**6**(8):e04709. DOI: 10.1016/j.heliyon.2020.e04709

[16] Basik AA, Sanglier J-J, Yeo CT, et al. Microbial degradation of rubber: Actinobacteria. Polymers. 2021;**13**(12):1989. DOI: 10.3390/polym13121989

[17] Ryan PG. In: Takada H, Karapanagioti H, editors. Ingestion of plastics by marine organisms, Hazardous Chemicals Associated with Plastics in the Marine Environment. The Handbook of Environmental Chemistry. Champions: Springer; 2016. p. 78. DOI: 10.1007/698_2016_21

[18] Pathak VM, Verma VK, Rawat BS, et al. Current status of pesticide effects on environment, human health and it's eco-friendly management as bioremediation: A comprehensive review. Frontiers in Microbiology. 2022;**13**:962619. DOI: 10.3389/fmicb.2022.962619

[19] Driscoll CT, Mason RP, Chan HM, et al. Mercury as a global pollutant: Sources, pathways, and effects. Environmental Science and Technology. 2013;**47**(10):4967-4983. DOI: 10.1021/es305071v

[20] Mohanan N, Montazer Z, Sharma PK, et al. Microbial and enzymatic degradation of synthetic plastics. Frontiers in Microbiology. 2020;**11**:580709. DOI: 10.3389/fmicb.2020.580709

[21] Bahl S, Dolma J, Singh JJ, et al. Biodegradation of plastics: A state of the art review. Materials Today: Proceedings. 2021;**39**(1):31-34. DOI: 10.1016/j.matpr.2020.06.096

[22] Basak SN, Meena SS. Microbial biodegradation of plastics: Challenges, opportunities, and a critical perspective. Frontiers in Environmental Science and Engineering. 2022;**16**:161. DOI: 10.1007/s11783-022-1596-6

[23] Srikanth M, Sandeep TSRS, Sucharitha K, et al. Biodegradation of plastic polymers by fungi: A brief review. Bioresources and Bioprocess. 2022;**9**:42. DOI: 10.1186/s40643-022-00532-4

[24] Knott BC, Erickson E, Allen MD, et al. Characterization and engineering of a two-enzyme system for plastics depolymerization. Proceedings of National Academy of Science USA. 2020;**117**(41):25476-25485. DOI: 10.1073/pnas.2006753117

[25] Quartinello F, Kremser K, Schoen H, et al. Together is better: The rumen microbial community as biological toolbox for degradation of synthetic polyesters. Frontiers in Bioengineering and Biotechnology. 2021;**9**:684459. DOI: 10.3389/fbioe.2021.684459

[26] Bhusare BP, Zambare VP, Jaweed TH, et al. Ecotoxicological impact of plastic waste on marine flora. In: Shahnawaz M, Sangale MK, Daochen Z, et al., editors. Impact of Plastic Waste on the Marine Biota. Singapore: Springer; 2022. pp. 257-286. DOI: 10.1007/978-981-16-5403-9_14

[27] Samir A, Ashour FH, Hakim AAA, et al. Recent advances in biodegradable polymers for sustainable applications. NPJ Material Degradation. 2022;**6**:68. DOI: 10.1038/s41529-022-00277-7

Chapter 2

The Strategy and Future of Biotechnology in Protecting the Global Environment

Naofumi Shiomi

Abstract

Global warming is accelerating, and the average global temperature is projected to rise from 3.5 to 5.7°C by the end of this century. Therefore, there is a strong possibility that we will soon experience frequent global-scale abnormal weather events and severe water and food shortages. To avoid such crises, three issues must be urgently addressed: reduction of CO_2 emissions, securing of energy sources that can replace fossil fuels, and securing of groundwater and food supplies. In this introductory chapter, we first discuss the development of new biotechnology processes such as CO_2 sequestration by algae, biofuels, and biopolymers. Biofuels and biopolymers, in particular, will soon play an important role as alternatives to scarce fossil fuels. In addition, bioremediation technologies for widespread groundwater and soil contamination are discussed. Novel bioremediation technologies, such as gene editing and the use of artificial enzymes, have the potential to dramatically improve bioremediation throughput. This new biotechnological approach to the environment will be a decisive factor in ensuring food and beverage safety.

Keywords: CO_2 capture, biofuel, biopolymer, bioremediation, global warming

1. Introduction

The Earth's climate has been undergoing significant changes over the past centuries, and the impact of these alterations is becoming increasingly evident. According to the Fourth Assessment Report of the United Nations Intergovernmental Panel on Climate Change, the concentration of carbon dioxide (CO_2) in the atmosphere has surged by more than 40% since pre-industrial times (1750). It is estimated to surpass the level of 400 parts per million (ppm) by 2013. This rise in CO_2 and methane, the two primary greenhouse gases, has led to a 1.09°C increase in the average global temperature for over 260 years. Notably, the rate of temperature increase is accelerating rapidly, and it is projected that the average global temperature will further increase by 1.5°C by 2027 and 3.3–5.7°C by the end of this century [1].

With the intensification of global warming, the world is grappling with an increasing frequency of extreme weather events, such as storms and heavy rainfall, and the consequences are already taking a toll on the global economy. The expansion of seawater, induced by global warming, has led to a 16-cm rise in the world's average sea level within the last century. This phenomenon has raised concerns about the imminent inundation of the South Pacific Islands and the potential submergence of the entire landmasses in some regions. Moreover, the regions that rely on glaciers and melting snow as their water source, covering one-sixth of the world's population, will face water scarcity as these ice formations continue to vanish due to global warming. Mountainous areas, which serve as the origins of major rivers, will also experience similar challenges. The impact of global warming on land and marine ecosystems is immense, causing extensive damage to plants, animals, and marine life. This destruction leads to the disruption of ecosystems, exacerbating the problem further. Notably, phytoplankton, a crucial CO_2 sink responsible for absorbing 31% of CO_2 emissions, is severely affected, accelerating global warming. Considering the current trajectory, it is evident that by the end of this century, the average global temperature will rise by 3.3–5.7°C, pushing Earth into a critical situation [2].

In addition to the direct impacts of global warming, water shortages have emerged as a severe concern, exacerbated by various factors beyond climate change [3]. The extensive implementation of large-scale projects that utilize rivers for irrigation in agricultural lands, along with population growth and rapid global economic development, has significantly increased the demand for water for domestic and industrial purposes, resulting in severe shortages in many regions. For instance, Egypt suffers from severe water scarcity as countries upstream of the Nile River consume disproportionate amounts of water and contribute to pollution [4]. According to a report by the United Nations in 2007, approximately 660 million people in 30–40 countries face extreme water shortages, whereas severe water shortages affect about 1.1 billion people. Tragically, 1.8 million children succumb to diseases caused by drinking contaminated water every year. Additionally, the combination of water scarcity and pollution in these countries adversely impacts crop growth, leading to water shortages and food scarcity, potentially escalating into international conflicts over vital resources in the future.

Given these pressing concerns, it is highly probable that humanity may soon encounter a time when only specific regions on Earth can sustain life. To ensure a sustainable society, the world must mobilize science and technology to address critical issues, such as reducing CO_2 emissions, developing renewable energy sources as alternatives to fossil fuels, and remediating pollution to secure safe food and drinking water. This chapter outlines the crucial role that biotechnology can play in confronting these challenges and providing potential solutions (**Figure 1**).

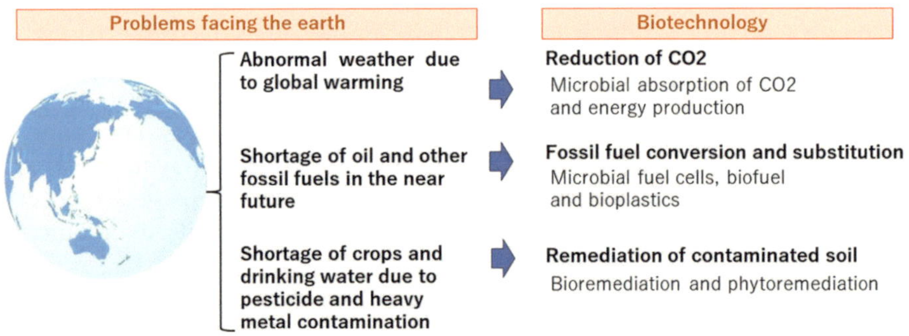

Figure 1.
Problems facing the earth and strategies for solving them using biotechnology.

2. Biotechnology strategies for global warming prevention and renewable energy development

2.1 CO_2 reduction

To address the urgent challenge of global warming, it is crucial to reduce atmospheric CO_2 levels. In this endeavor, countries worldwide have united under the "Paris Agreement," committed to mitigating CO_2 emissions. For instance, the Japanese government has taken a significant step by announcing the "2050 Carbon Neutral Declaration," setting the ambitious goal of achieving zero carbon emissions. To further prevent atmospheric CO_2 increase, global CO_2 emissions must be reduced to 50% by 2050 from 2006 levels, with the ultimate target of achieving a 100% reduction by 2075. However, despite the critical situation, major contributors to CO_2 emissions, such as China and the United States, are yet to embrace CO_2 capture technologies, emphasizing the need for innovative and effective technological advancements.

The most promising method of reducing atmospheric CO_2 is the "carbon capture and storage (CCS) process," in which atmospheric CO_2 is compressed and stored in liquid form at an underground site or the bottom of the deep ocean [5]. Although this method can significantly reduce the concentration of CO_2 in the atmosphere, its high cost is a major problem [6, 7]. Currently, converting CO_2 into value-added fuels instead of treating it as waste is considered economical, and efforts are being made to achieve this. For example, several studies have been conducted to convert CO_2 into fuels through electrochemical methods, and solar-electric lands have been developed to convert CO_2 into hydrocarbon fuels using only sunlight [8].

CO_2 capture using microorganisms is attracting attention as an environmentally friendly method [9]. Microalgae, in particular, are the best CO_2 capture sources compared with higher plants primarily due to their exceptional solar energy yield and year-round cultivability. Because the production cost of microalgae for CO_2 capture must be less than 500 t/ha/year, efforts are being made to reduce this cost. For example, there is a lower production cost in conjunction with the production of various useful substances. Using genetically modified microalgae, bioconversion processes from CO_2 to biofuels, polyhydroxybutyrate, fatty acid ethyl esters, and other useful substances have been considered [10, 11]. Various novel production processes have also been devised, including the direct use of exhaust gas from thermal power plants or wastewater treatment facilities as sources of carbon dioxide gas, the development of photobioreactors suitable for large-scale outdoor cultivation, and microbial mineral carbonization in combination with metal recovery [12], which are considered key to success.

2.2 Fossil fuel conservation and substitution

The rapid increase in energy consumption has raised concerns regarding the depletion of fossil fuels, such as oil and gas. According to the Hubbert Peak Theory, oil resources are expected to be severely depleted within the next 40 years [13]. It is also estimated that currently mined oil reserves will be exhausted in 42 years and natural gas reserves in 60 years. In reality, fossil fuel depletion is unlikely to occur because there are still unexploited fossil fuels on Earth, but the prices of newly mined oil and gas will be much higher than those of today. Furthermore, the consumption of fossil fuels is a major source of CO_2 emissions; thus, fossil fuel consumption must be reduced. Therefore, shifting from fossil fuels to sustainable renewable energy is important for securing energy and preventing global warming [14]. Here, we introduce renewable energy initiatives from a biotechnological perspective.

Biofuels are used as renewable alternative fuels [15, 16]. Ethanol and biodiesel, produced from food and animal feed such as corn, soybeans, rapeseed, or used cooking oil, have been developed as first-generation biofuels. They are already used in transportation as substitutes for conventional fuels, and their production continues to increase. However, producing these biofuels requires large tracts of farmland, which is undesirable because it pressurizes food production. Therefore, second-generation biofuels have been designed using inedible lignocellulosic feedstocks accumulated as agricultural waste. However, this method has not yet been put to practical use because of issues related to the stability of the feedstock supply and its preprocessing.

Third-generation biofuels are currently being developed via transesterification and hydrogenation of oil produced by microalgae [17]. As mentioned, microalgae are ideal for biofuel production and global warming mitigation because of their high CO_2 fixation capacity and, at the same time, high lipid content. Microalgae are also superior in that they can produce various biofuels, such as hydrogen and biodiesel by transesterification, bio-oil, and bioethanol, but their production costs are higher than those of fossil fuels and biodiesel. Currently, genetically modified algae and algae improved by gene editing using CRISPR/Cas9 have been developed to increase oil accumulation. Wastewater treatment and the use of photochemical and electric fuels are also being investigated [18, 19].

Cyanobacteria, considered as photosynthetic microorganisms, have been genetically modified to produce biofuels [20]. Cyanobacteria produce biofuels from the carbon produced by photosynthesis using CO_2 and water. For example, genetic modification has been reported to improve the production of bioethanol, isobutanol, isoprene, and fatty acids that are attracting attention as alternatives to gasoline [21], and strains that secrete biofuels extracellularly have also been created through genetic modification. If the use of genetically modified algae and cyanobacteria is accepted in biofuel production, they may be used as renewable alternative fuels.

To reduce CO_2 emissions and fossil fuel consumption, gasoline-powered vehicles are being replaced by electric vehicles (EVs). The European Union (EU) has effectively banned the sale of gasoline-powered vehicles, including hybrid vehicles, for 35 years and has decided to shift to EVs and fuel-cell vehicles. However, securing sufficient electricity from hydroelectric and wind power generation is difficult and replacing all cars with EVs will require generating vast amounts of new electricity. For example, Norway has become a leading EV country with revenues generated by its abundant oil and natural gas, but as a result, the fossil fuels that Norway exports emit CO_2, which has not led to a reduction in CO_2 emissions. In the future, it will be necessary to face an electricity shortage by generating electricity from solar and fuel cells in each household instead of supplying electricity from electric power companies, such as in the past.

Hydrogen fuel cell is one of the promising fuel cells [22, 23]. Hydrogen fuel cells generate electricity from chemical energy using hydrogen and oxygen and have many advantages, such as almost zero emission of CO_2 and environmental pollutants, low vibration and noise, easy installation inside buildings and in urban areas, and high energy efficiency using waste heat. Hydrogen fuel cells can also be used as fuel for cars and trucks. Hydrogen-fueled vehicles are being developed in China and Japan. Various fuels, such as natural gas and methanol, can be used to extract hydrogen, but the challenge is to produce hydrogen at a low cost. Although the production capacity is not yet sufficient, methods using photosynthetic microalgae and cyanobacteria have been investigated to produce hydrogen gas [24, 25]. Microalgae that absorb CO_2 while producing hydrogen are attracting particular attention because natural gas emits CO_2 during hydrogen production.

In contrast, the development of microbial fuel cells (MFCs) is also underway [26, 27] MFCs are bioreactors that convert organic compounds into electrical energy under anaerobic conditions through microbial catalysis and convert biomass into electricity, such as wood and food waste as well as wastewater. It is a desirable renewable energy source. Currently, MFCs have not been put to practical use because of their high cost, low expression rate, and lack of scaled-up capabilities. However, in recent years, with the use of molecular tools and genetic systems of microorganisms combined with genetic manipulation and electrochemistry, improvements in electric fields and a new style of MFC using wastewater [28, 29], the amount and stability of power generation have rapidly improved and future developments are expected.

2.3 Substitution of petroleum-based polymers

The CO_2 emitted by the production and incineration of plastics was about 850 million tons in 2019, whereas open-air combustion may emit 1 Gt [30]. In addition, newly initiated industrial projects continue to increase the production capacity of petrochemical plastics with a potential CO2 emission of 2.8 Gt by 2050 (10–15% of the global carbon budget). To reduce petroleum consumption and CO_2 emissions simultaneously, the consumption of petrochemical products, especially plastics, must be reduced; and the recycling and reuse of plastics must be implemented. Currently, PET bottles are being actively recycled, and plastic cups and other plastic products are being regulated in Japan. The substitution of paper products is progressing but not sufficiently. The dependence on paper as a substitute will lead to the deforestation of forests including mangroves, and the resulting sharp decline in the number of plants worldwide will spur an increase in atmospheric CO_2 emissions.

Bioplastics have attracted attention as alternatives to petroleum-based polymers [21, 31, 32]. Bioplastics are bio-based, biodegradable, or a combination of both. In addition to synthesizing new polymers, "drop-in polymers," which are substitutes for various petroleum-based polymers, can also be produced from biomass-derived carboxylic acids, alcohols, vinyl monomers, and amides. For example, the aliphatic homopolymer PLA can be produced by the polycondensation of lactic acid from fermentation (or ring-opening polymerization of lactide), with an annual production capacity of over 250,000 tons [33]; PHA is a polymer that can be synthesized by bacteria and accumulates in high concentrations in the cells, and the production of polymers from biomass is a promising alternative to petroleum-based polymers. It is expected to be produced at 100,000 t/year within the next few years [34]. In addition, PET is produced by polycondensation of the bio-derived monomer monoethylene glycol with 2,5-furandicarboxylic acid, and bioPET is obtained synthetically or microbially from biomass. PET production is expected to increase dramatically in the future, owing to the progress of PET recycling [34]. Although biopolymers are used in significantly smaller quantities than petroleum-derived polymers, they are expected to play an increasingly important role in reducing fossil fuel consumption.

3. Biotechnology strategies for global environmental restoration

3.1 Soil contamination resulting in severe food and drinking water scarcity

The excessive utilization of persistent organic pollutants (POPs) has severely damaged soil and groundwater. In the 1980s, these pollutants were widely employed as pesticides. Because of their prolonged persistence in the environment, they contaminated groundwater and have accumulated in birds and fish. To address this issue, the 2001 Stockholm Convention prohibited the production and use of POPs. Following that, organophosphorus pesticides (OPPs), which exhibit a much shorter persistence in soil than POPs, were banned in the EU in the 2000s following the reports of contamination in drinking water. Triazine herbicides developed in the 1970s were found to have endocrine-disrupting effects on frogs. The EU established an upper limit for their field use, whereas the U.S. Environmental Protection Agency legally regulated the maximum amount (3 ppm) of atrazine in drinking water. Despite these efforts, the excessive use of pesticides to improve crop yields [35] continues, particularly in Asia, Africa, and South America, exacerbating food concerns. Additionally, some areas still employ the stocks of POPs produced before the implementation of the Stockholm Convention.

Soil contamination by toxic metals is also becoming increasingly serious. For example, arsenic (As), because of the widespread use of Ga-As and Se-As semiconductors in consumer electronics, is produced in the mining industry; and large amounts of crude ore, sediments, and wastewater containing As have contaminated soil. High concentrations of As have been reported in groundwater in Asian countries, such as Vietnam, Thailand, India, and China [36–38]. In the Indian state of West Bengal, approximately 8 million people are at risk of arsenic poisoning. Other poisoning cases include As and fluorine in low-quality coal fuel and in some areas of China's Sichuan and Guizhou provinces, where coal containing very high concentrations of fluorine (500 mg/kg) is used, leading to coal-fueled fluorine poisoning in approximately 20 million people.

Pollution by toxic heavy metals, such as lead, cadmium, mercury, and hexavalent chromium, has also caused health problems [39, 40]. For example, the production of lead-acid batteries has increased sharply owing to the increasing demand for automobiles; and the direct discharge of exhaust gases and wastewater containing high concentrations of lead from many metallurgical and mining operations with inadequate pretreatment, low-quality ore, or used sludge left on the soil has caused severe lead poisoning in areas surrounding these operations. Cadmium is mainly used in Li-Cd batteries in consumer electronics, but because the recycling rate of Li-Cd batteries is as low as 20%, cadmium poisoning has been reported in various areas and around mining sites.

In 2000, the EU promulgated the waste electrical and electronic equipment (WEEE) Directive, which mandated the recycling of end-of-life vehicles and strictly limited the use of hazardous metals [41]. The 2011 revised RoHS Directive (RoHS2) expanded the number of products subject to it to approximately 20,000 products. As a result of this strict directive, the recycling rate of lead-acid batteries exceeded 90%; and alternatives that did not contain toxic metals, such as lead-free solders, were developed. However, nearly 80% of WEEE is still disposed of, and technical solutions are required. The most effective measures to halt the progression of soil and groundwater contamination are to reduce the discharge of pollutants through strict legal regulations and treaties, but at the same time, the ever-widening contaminated soil and groundwater must be remediated as soon as possible. If this is not done, there is a certainty that food and drinking water shortages will occur in the future.

3.2 Soil remediation

Bioremediation is a promising technology that uses living organisms to remove harmful contaminants through degradation, adsorption, or absorption and has the advantages of being cost-effective and widely applicable compared with physicochemical methods [42]. For *in situ* methods, the bioaugmentation method that promotes decomposition by activating indigenous microorganisms with nutrients and oxygen has been primarily used rather than the biostimulation method that adds highly degradable microorganisms [43, 44]. This is because of concerns about the damage to the ecosystem caused by the introduction of nonindigenous microorganisms. However, biostimulation methods have limited restorative capabilities. Considering the wide variety of substances currently contaminating the soil and the wide range of contamination, bioaugmentation methods must be used for remediation in the future.

Phytoremediation, which involves the use of plants for restoration, is also expected. In the case of microorganisms, it is difficult to recover from their spread in the soil. In contrast, phytoremediation is a method in which contaminants near the soil surface are absorbed or adsorbed by plants; as contaminants can be removed from the soil together with the plants, phytoremediation is advantageous for the recovery of heavy metals that cannot be decomposed. Plants are susceptible to environmental influences, do not necessarily have a high remediation capacity, and are often not economically viable. To solve this problem, several plants that can effectively absorb and accumulate metal ions, called "hyperaccumulators," have been discovered [45]. For example, *Rinorea nicolifera*, found in the Philippines, can accumulate 18,000 ppm of nickel [46]. Furthermore, attempts have been made to reduce production costs by combining them with plants that produce useful substances, such as herb-producing plants.

Recently, enzyme-based bioremediation methods have attracted significant attention [47, 48]. Enzymes, such as oxidoreductase, laccase, hydrolase, and peroxidase, are used in bioremediation. Because they are proteins, they do not require the removal of accumulated biomass from the treated area; however, they are not suitable for the adsorption of heavy metals because they cannot be removed without immobilization. Therefore, various methods for enzyme immobilization have been investigated [48].

3.3 Biotechnological strategies for practical remediation

As mentioned earlier, wild strains of microorganisms and plants have not achieved economically viable remediation capabilities. For practical use, it is necessary to use strains with tens to hundreds of times higher decomposition capacities and excellent operability using the latest technologies, including genetic modifications. Another advantage of using recombinant microorganisms and plants is that they can provide the ability to decompose and absorb pollutants even under extreme climatic conditions. Widespread pollution occurs mainly in harsh climates, such as acidic, cold, and arid climates. However, microbial activity and plant growth are less resistant to extreme environments, which is a disadvantage of biological remediation processes. Therefore, it would be useful to utilize genetic technology to add the ability to maintain high activity in harsh climates. Many microorganisms that degrade and remove pollutants have been discovered [49]. For example, many microorganisms with improved degradation ability of triazine pesticides and organophosphorus herbicides and microorganisms with high absorption and adsorption ability for heavy metals have been discovered; their degradation pathways and absorption mechanisms have been clarified, and most degradation genes have been cloned.

Furthermore, new methods are being developed to identify and utilize enzymes from unknown organisms and artificial enzymes using new techniques. For example, metagenomics is a method used to identify the enzymes of unknown organisms. Ninety-nine percent of all microorganisms are unculturable, and their genetic information has been overlooked. Conversely, metagenomics makes it possible to extract the genetic information of unculturable microorganisms and dramatically increase the repertoire of enzymes. It is now possible to obtain genetic information on highly active enzymes and microorganisms in specific environments [50, 51]. For example, fewer than 20 nitrilases have been discovered using conventional screening, whereas more than 300 unknown nitrilases have been discovered in metagenomic libraries [52]. Using this method, metagenomic analysis is becoming essential for bioremediation, as new metallothionein families (environmental metallothioneins) have been found [53]; and metagenomic analysis is actively used for the removal of heavy metals, such as hexavalent chromium, which is becoming indispensable for bioremediation.

Molecular evolutionary engineering has facilitated the development of superior artificial enzymes [54]. Artificial enzymes with excellent activity and stability were produced after introducing mutation either by site-mutagenesis based on computational design or Et-PCR or DNA shuffling and repeating the selection process using DNA, RNA, or ribosomal display. For example, the thermostability of transglutaminase was improved 12-fold at 60°C by introducing saturating mutations [55]. In addition, computational design and site-mutagenesis methods have been used to create eight new enzymes that catalyze Kemp dephosphorylation reactions that are not found in nature [56]. The modification of enzymes for environmental remediation by molecular evolutionary engineering is still in its infancy, but it is expected to become an important area in the future.

CRISPR-Cas9 and other genome-editing technologies are also being actively used. CRISPR-Cas9 is a technology for silencing by cutting specific gene portions. However, further advances have led to the use of CRISPRa and CRISPRi to control gene expression, as well as CRISPR to rewrite gene sequences, such as BaseEditor or Target-AID. This novel technology is convenient for use in environmental restoration because (1) it can edit both genes, even if the chromosomes are diploid, making it excellent for gene editing in plants and other organisms; and (2) it is not part of a genetically modified organism, making it suitable for environmental restoration. Many genetically edited plants have been produced in recent years, and their decomposition and adsorption activities have increased. For example, gene editing has produced rice and corn with improved yields or plants that have been enhanced with carotene and other substances [57], and the application of plants for phytoremediation has rapidly expanded in recent years [58, 59].

In synthetic biology, attempts are being made to use these new technologies and databases to redesign and reconstruct microorganisms and create new biological systems (or reaction modules) with enhanced functions [60]. Optimized devices, or microorganisms carrying these devices, have been created by restructuring molecular tools and genetic frameworks to degrade pollutants. Wang *et al*. created a microorganism in which a phenol hydroxylase gene and seven catecholamine-degrading genes were incorporated into *Escherichia coli* [61]. This microorganism uses phenol as the carbon source in wastewater contaminated with crude oil. Dueber et al. increased the reaction rate nearly 89-fold by creating a synthetic protein scaffold (module) that closely aligned the binding domains with the enzymes involved in the pathway, mimicking the substrate tunneling effect [62].

Consortia are also important for improving bioremediation capacity. Plants and microorganisms, such as rhizobacteria, live in natural settings, helping each other; and there is excellent synergy, especially in the case of toxic metal ion adsorption [63, 64]. Although many microorganisms exhibit a high resistance and adsorption capacity for heavy metals, removing metal-adsorbed microorganisms from the site is complex. The removal of metal-adsorbed plants from soil is easy; however, the resistance of plants to heavy metals is lower than that of microorganisms, and they cannot adsorb contaminants below a depth of 1–2 m. Therefore, a combination of rhizobacteria and plants would enable effective remediation. Plants undergo apoptosis when the water content is insufficient. If microorganisms that secrete polymers, such as polyglutamic acid and chondroitin, are combined with plants, they can be used as soil humectants, and some plants can grow even in dry climates. Moreover, aloe can grow with little water, accumulate a large amount of water components in its leaves, and keep growing new leaves without withering [65], suggesting that aloe can be used as a host plant for genetic manipulation for more effective remediation.

Nanoremediation, a combination of nanotechnology and bioremediation, has attracted attention as a novel technology [66]. Nanoparticles (NPs) are useful tools for remediating contaminants because of their small size (1–100 nm), large specific surface area, and reactivity for adsorption, redox reactions, and precipitation. The combination of nanoparticles and microorganisms can promote purification efficiency. For example, in nano-phytoremediation, which combines NPs with plants, the NPs mitigate the toxicity of metal contaminants and promote plant growth. Consequently, the removal rate of toxic substances improved. However, the safety of NPs released into the ecosystem has not yet been well studied and warrants further research. However, nanoparticles are toxic for humans and for certain animals due to their small size (nano size). Especially in cases of inhalation, nanoparticles (NP) mixed with oxygen or air can cause serious life-threatening problems. Furthermore, during breathing, the lungs, brain, and cardiovascular neural system are exposed to serious life-threatening illnesses, such as pneumonia, chronic cough, inflammation of the trachea, lung diseases, heart diseases as well as mental disorders and disorders of the nervous system. Therefore, when working with NPs, it is necessary to observe the maximum safety and to be well informed about the toxicity of nanoparticles (NP) and the potential danger to the human body.

4. Conclusions

The global environment is deteriorating rapidly. To ensure a sustainable society, there is an urgent need to prevent global warming by reducing greenhouse gases, developing renewable energy sources to replace fossil fuels, and securing drinking water and food supplies by remediating soil and groundwater contamination. It is no exaggeration to say that if these measures are not accomplished, humanity will face a crisis by the end of this century. These issues must be addressed by combining various scientific technologies; biotechnology is also important. The production of biofuels, bioplastics, and microbial batteries using microorganisms can reduce CO_2 emissions and provide renewable energy. The remediation of soil and groundwater contamination using microorganisms and plants can increase the availability of drinking water. Biotechnology is expected to play a significant role in the future.

However, several problems to overcome in the future are still present. One is expensive running costs because of insufficient production and degradation capacities of microorganisms. To solve this problem, it is necessary to construct prominent microorganisms by the use of genetic modification, gene editing, and molecular evolutionary engineering and to build optimal processes. Another problem is the difficulty of using genetically modified organisms in the environment. The legal regulation prevents the widespread use of fuel production by recombinant algae and bioremediation with recombinant microorganisms or plants. Although, we must carefully consider the impact on ecosystems; I think that the active use of genetically modified bacteria and plants in the environment cannot be avoided because it may be a final option to realize a sustainable society in the future.

Author details

Naofumi Shiomi
School of Human Sciences, Kobe College, Japan

*Address all correspondence to: n-shiomi@mail.kobe-c.ac.jp

References

[1] Matt MGM. Global Warming Set to Break Key 1.5°C Limit for First Time. BBC NEWS COP27. London: British Broadcasting Corporation; 2023

[2] United Nations Press. Only 11 Years Left to Prevent Irreversible Damage from Climate Change, Speakers Warn during General Assembly High-Level Meeting. New York: United Nations Press; 2019. p. GA12131. Available from: https://press.un.org/en/2019/ga12131.doc.htm

[3] Reports of UNICEF. Thiraring for a Future: Water and Chirdren in a Changing Climate. New York: Reports of UNICEF; 2017

[4] Omar MEDM, Moussa AMA. Water management in Egypt for facing the future challenges. Journal of Advanced Research. 2016;7:403. DOI: 10.1016/j.jare.2016.02.005

[5] Shreyash N, Sonker M, Bajpai S, et al. The review of carbon capture-storage technologies and developing fuel cells for enhancing utilization. Energies. 2021;**14**:4978. DOI: 10.3390/en14164978

[6] Rubin ES, Davison JE, Herzog HJ. The cost of CO_2 capture and storage. International Journal of Greenhouse Gas Control. 2015;**40**:378. DOI: 10.1016/j.ijggc.2015.05.018

[7] Lilliestam J, Bielicki JM, Patt AG. Comparing carbon capture and storage (CCS) with concentrating. Solar power (CSP): Potentials, costs, risks, and barriers. Energy Policy. 2012;**47**:447. DOI: 10.1016/j.enpol.2012.05.020

[8] Asadi M, Kim K, Liu C, et al. Nanostructured transition metal dichalcogenide electrocatalysts for CO_2 reduction in ionic liquid. Science. 2016;**353**:467. DOI: 10.1126/science.aaf4767

[9] Jajesniak P, Ali HO, Wong TS. Carbon dioxide capture and utilization using biological systems: Opportunities and challenges. Journal of Bioprocessing and Biotechniques. 2014;**4**:3. DOI: 10.4172/2155-9821.1000155

[10] Morales M, Sánchez L, Revah S. The impact of environmental factors on carbon dioxide fixation by microalgae. FEMS Microbiology Letters. 2018;**365**:fnx262. DOI: 10.1093/femsle/fnx262

[11] Branduardi P, Sauer M. Microbial carbon dioxide fixation: New tricks for an old game. FEMS Microbiology Letters. 2018;**365**:fnx269. DOI: 10.1093/femsle/fnx269

[12] McCutcheon J, Power IM. Microbially mediated carbon dioxide removal for sustainable mining. PLoS Biology. 2023;**21**:e3002026. DOI: 10.1371/journal.pbio.3002026

[13] Deming D, Deming DM. King Hubbert and the rise and fall of peak oil theory. AAPG Bulletin. 2023;**107**:851. DOI: 10.1306/03202322131

[14] Sanjukta S, Mahapatra M. Biotechnological intervention in global warming: Climate change and water crisis. In: Thokchom B et al., editors. Water Conservation in Era of Global Climate Change. Amsterdam: Elsevier; 2021. p. 315. DOI: 10.1016/B978-0-12-820200-5.00010-5

[15] Jeswani HK, Chilvers A, Azapagic A. Environmental sustainability of biofuels: A review. Proceedings of the Royal Society of London Series A. 2020;**467**:2243. DOI: 10.1098/rspa.2020.0351

[16] Mahapatra S, Kumar D, Singh B, Sachan PK. Biofuels and their sources

of production: A review on cleaner sustainable alternative against conventional fuel, in the framework of the food and energy nexus. Energy Nexus. 2021;**4**:100036. DOI: 10.1016/j.nexus.2021.100036

[17] Wood DA. Microalgae to biodiesel - review of recent progress. Bioresource Technology Reports. 2021;**14**:100665. DOI: 10.1016/j.biteb.2021.100665

[18] Abdullah B, Syed Muhammad SAF, Shokravi Z, et al. Fourth generation biofuel: A review on risks and mitigation strategies. Renewable and Sustainable Energy Reviews. 2019;**107**:37. DOI: 10.1016/j.rser.2019.02.018

[19] Khoo KS, Ahmad I, Chew KW, Iwamoto K, Bhatnagar A, Show PL. Enhanced microalgal lipid production for biofuel using different strategies including genetic modification of microalgae: A review. Progress in Energy and Combustion Science. 2023;**96**:101071. DOI: 10.1016/j.pecs.2023.101071

[20] Farrokh P, Sheikhpour M, Kasaeian A, Asadi H, Bavandi R. Cyanobacteria as an eco-friendly resource for biofuel production: A critical review. Biotechnology Progress. 2019;**35**:e2835. DOI: 10.1002/btpr.2835

[21] Rosenboom JG, Langer R, Traverso G. Bioplastics for a circular economy. Nature Reviews. Materials. 2022;**7**:117. DOI: 10.1038/s41578-021-00407-8

[22] Singla MK, Nijhawan P, Oberoi AS. Hydrogen fuel and fuel cell technology for cleaner future: A review. Environmental Science and Pollution Research International. 2021;**28**:15607. DOI: 10.1007/s11356-020-12231-8

[23] Staffell I, Scamman D, Velazquez Abad AD, et al. The role of hydrogen and fuel cells in the global energy system. Energy and Environmental Science. 2019;**12**:463. DOI: 10.1039/C8EE01157E

[24] Ahmed SF, Rafa N, Mofijur M, et al. Biohydrogen production from biomass sources: Metabolic pathways and economic analysis. Frontiers in Energy Research. 2021;**9**:753878. DOI: 10.3389/fenrg.2021.753878

[25] Tran QN, Kim IT. A review of biohydrogen production from Saccharina japonica. Fermentation. 2023;**9**:242. DOI: 10.3390/fermentation9030242

[26] Azuma M, Ojima Y. Catalyst development of microbial fuel cells for renewable-energy production. In: Shiomi N, editor. Current Topics in Biochemical Engineering. London, UK, Rejeca: InTech; 2018. DOI: 10.5772/intechopen.81442

[27] Vishwanathan AS. Microbial fuel cells: A comprehensive review for beginners. 3 Biotech. 2021;**11**:248. DOI: 10.1007/s13205-021-02802-y

[28] Prathiba S, Kumar PS, Vo D-VN. Recent advancements in microbial fuel cells: A review on its electron transfer mechanisms, microbial community, types of substrates and design for bio-electrochemical treatment. Chemosphere. 2022;**286**:131856. DOI: 10.1016/j.chemosphere.2021.131856

[29] Luo J, Tian W, Jin H, et al. Recent advances in microbial fuel cells: A review on the identification technology, molecular tool and improvement strategy of electricigens. Current Opinion in Electrochemistry. 2023;**37**:101187. DOI: 10.1016/j.coelec.2022.101187

[30] United Nations Framework Convention on Climate Change (UNFCCC). Report of the Conference of the Parties on its Twenty-First Session,

Held in Paris from 30 November to 13 December 2015. Paris: UNFCCC; 2016

[31] Saharan R, Kharb J. Exploration of bioplastics: A review. Oriental Journal of Chemistry. 2022;**38**:840. DOI: 10.13005/ojc/380403

[32] Flórez M, Cazón P, Vázquez M. Selected biopolymers' processing and their applications: A review. Polymers. 2023;**15**:641. DOI: 10.3390/polym15030641

[33] Ashothaman A, Sudha J, Senthilkumar N. A comprehensive review on biodegradable polylactic acid polymer matrix composite material reinforced with synthetic and natural fibers. Materials Today: Proceedings. 2023;**80**:2829. DOI: 10.1016/j.matpr.2021.07.047

[34] Mozejko-Ciesielska J, Kumar P, Lemos PC, Cui Y. Editorial: Advances and trends in polyhydroxyalkanoate (PHA) biopolymer production. Frontiers in Bioengineering and Biotechnology. 2022;**10**:873250. DOI: 10.3389/fbioe.2022.873250

[35] Khan S, Naushad M, Lima EC, Zhang S, Shaheen SM, Rinklebe J. Global soil pollution by toxic elements: Current status and future perspectives on the risk assessment and remediation strategies - a review. Journal of Hazardous Materials. 2021;**417**:126039. DOI: 10.1016/j.jhazmat.2021.126039

[36] Shiomi N. Introductory chapter: Serious pollution of soil and groundwater and the necessity of bioremediation. In: Shiomi N, editor. Advances in Bioremediation and Phytoremediation. London, UK, Rijeca: InTech; 2018. pp. 1-17. DOI: 10.5772/intechopen.74403

[37] Shankar S, Shanker U, Shikha. Arsenic contamination of groundwater: A review of sources, prevalence, health risks, and strategies for mitigation. The Scientific World Journal. 2014;**2014**:304524. DOI: 10.1155/2014/304524

[38] Rodríguez-Lado L, Sun G, Berg M, et al. Groundwater arsenic contamination throughout China. Science. 2013;**341**:866. DOI: 10.1126/science.1237484

[39] Shiomi N. An assessment of the causes of lead pollution and the efficiency of bioremediation by plants and microorganisms. In: Shiomi N, editor. Advances in Bioremediation of Wastewater and Polluted Soil. London, UK, Rijeca: IntechOpen; 2015. pp. 247-174. DOI: 10.5772/60802

[40] Kubier A, Wilkin RT, Pichler T. Cadmium in soils and groundwater: A review. Applied Geochemistry. 2019;**108**:1. DOI: 10.1016/j.apgeochem.2019.104388

[41] Health and safety Exsective. Waste Electrical and Electronic Equipment recycling (WEEE). Available from: https://www.hse.gov.uk/waste/waste-electrical.htm

[42] Sharma I. Bioremediation techniques for polluted environment: Concept, advantages, limitations, and prospects. In: Murillo-Tovar MA, HAS N, Saeid A, editors. Trace Metals in the Environment - New Approaches and Recent Advances. London, UK, Rijeca: IntechOpen; 2021. pp. 1-16. DOI: 10.5772/intechopen.90453

[43] Winardi SES, Haryono E, Sudrajat. Research roadmap of bioremediation: Review of in situ method on land bioremediation. Journal of Physics: Conference Series. 2018;**1175**:012130. DOI: 10.1088/1742-6596/1175/1/012130

[44] Sharma P, Pandey AK, Kim S-H, Singh SP, Chaturvedi P, Varjani S.

Critical review on microbial community during in-situ bioremediation of heavy metals from industrial wastewater. Environmental Technology and Innovation. 2021;**24**:101826. DOI: 10.1016/j.eti.2021.101826

[45] Padmavathiamma PK, Li LY. Phytoremediation technology: Hyper-accumulation metals in plants. Water, Air, and Soil Pollution. 2007;**184**:105. DOI: 10.1007/s11270-007-9401-5

[46] Fernando ES, Quimado MO, Doronila AI. Rinorea niccolifera (Violaceae), a new, nickel hyperaccumulating species from Luzon Island, Philippines. Phyto Keys. 2014;**37**:1-13. DOI: 10.3897/phytokeys.37.7136

[47] Karigar CS, Rao SS. Role of microbial enzymes in the bioremediation of pollutants: A review. Enzyme Research. 2011;**2011**:805187. DOI: 10.4061/2011/805187

[48] Somu P, Narayanasamy S, Gomez LA, Rajendran S, Lee YR, Balakrishnan D. Immobilization of enzymes for bioremediation: A future remedial and mitigating strategy. Environmental Research. 2022;**212**:113411. DOI: 10.1016/j.envres.2022.113411

[49] Azad MAK, Amin L, Sidik NM. Genetically engineered organisms for bioremediation of pollutants in contaminated sites. Chinese Science Bulletin. 2014;**59**:703. DOI: 10.1007/s11434-013-0058-8

[50] Kumar V, Bilal M, Shahi SK, Garg V, editors. Metagenomics to Bioremediation: Applications, Cutting Edge Tools, and Future Outlook. Amsterdam: Elsevier; 2022

[51] Devarapalli P, Kumavath RN. Metagenomics -a technological drift in bioremediation. In: Shiomi N, editor. Advances in Bioremediation of Wastewater and Polluted Soil. Rejeca: London, UKIntechOpen; 2014. pp. 74-91. DOI: 10.5772/60749

[52] Robertson DE, Chaplin JA, DeSantis G, et al. Exploring nitrilase sequence space for enantioselective catalysis. Applied and Environmental Microbiology. 2004;**70**:2429. DOI: 10.1128/AEM.70.4.2429-2436.2004

[53] Ziller A, Yadav RK, Capdevila M, et al. Metagenomics analysis reveals a new metallothionein family: Sequence and metal-binding features of new environmental cysteine-rich proteins. Journal of Inorganic Biochemistry. 2017;**167**:1. DOI: 10.1016/j.jinorgbio.2016.11.017

[54] Shiomi N. Introductory chapter: Artificial enzyme produced by directed evolution technology. In: Shiomi N, editor. Current Topics in Biochemical Engineering. Rijeca: London, UKIntechOpen; 2019. pp. 1-10. DOI: 10.5772/intechopen.85738

[55] Buettner K, Hertel TC, Pietzsch M. Increased thermostability of microbial transglutaminase by combination of several hot spots evolved by random and saturation mutagenesis. Amino Acids. 2012;**42**(2-3):987-996. DOI: 10.1007/s00726-011-1015-y

[56] Röthlisberger D, Khersonsky O, Wollacott AM, et al. Kemp elimination catalysis by computational enzyme design. Nature. 2008;**453**:190. DOI: 10.1038/nature06879

[57] Shan Q, Wang Y, Li J, et al. Targeted genome modification of crop plants using a CRISPR-CAS system. Nature Biotechnology. 2013;**31**:686. DOI: 10.1038/nbt.2650

[58] Venegas-Rioseco J, Ginocchio R, Ortiz-Calderón C. Increase in phytoextraction potential by genome editing and transformation: A review. Plants. 2021;**11**:86. DOI: 10.3390/plants11010086

[59] Liu D-F, Li W-W. Genome editing techniques promise new breakthroughs in water environmental microbial biotechnologies. ACS ES&T Water. 2021;**1**:745. DOI: 10.1021/acsestwater.0c00276

[60] Jiménez-Díaz V, Pedroza-Rodríguez AM, Ramos-Monroy O, Castillo-Carvajal LC. Synthetic biology: A new era in hydrocarbon bioremediation. PRO. 2022;**10**:712. DOI: 10.3390/pr10040712

[61] Wang B, Xu J, Gao J, et al. Construction of an Escherichia coli strain to degrade phenol completely with two modified metabolic modules. Journal of Hazardous Materials. 2019;**373**:29. DOI: 10.1016/j.jhazmat.2019.03.055

[62] Dueber JE, Wu GC, Malmirchegini GR, et al. Synthetic protein scaffolds provide modular control over metabolic flux. Nature Biotechnology. 2009;**27**:753. DOI: 10.1038/nbt.1557

[63] Raklami A, Meddich A, Oufdou K, Baslam M. Plants—Microorganisms-based bioremediation for heavy metal cleanup: Recent developments, phytoremediation techniques, regulation mechanisms, and molecular responses. International Journal of Molecular Sciences. 2022;**23**:5031. DOI: 10.3390/ijms23095031

[64] Saeed Q, Xiukang W, Haider FU, et al. Rhizosphere bacteria in plant growth promotion, biocontrol, and bioremediation of contaminated sites: A comprehensive review of effects and mechanisms. International Journal of Molecular Sciences. 2021;**22**:10529. DOI: 10.3390/ijms221910529

[65] Katubi KM, Amari A, Harharah HN, Eldirderi MM, Tahoon MA, Ben RF. Aloe vera as promising material for water treatment: A review. PRO. 2021;**9**:782. DOI: 10.3390/pr9050782

[66] Rajput VD, Minkina T, Upadhyay SK, et al. Nanotechnology in the restoration of polluted soil. Nanomaterials. 2022;**12**:769. DOI: 10.3390/nano12050769

Chapter 3

Soil Treatment Technologies through Bioremediation

Ioana Stanciu

Abstract

Bioremediation includes processes such as bioventing, bioaugmentation, phytoremediation, biopiles, and composting. In this chapter, we details the characteristics, utilization and operating conditions of each process. Bioremediation is understood, according to the general definition, as the use of living organisms (microorganisms, plants, etc.) to improve and restore the ecological condition of a polluted or degraded substrate (area, land, aquifer, etc.) to better, favorable quality parameters life, harmless, non-polluting or to return it to its previous state. Soil treatment technologies through bioremediation include two types of treatments: in situ biological treatments (bioventilation, bioaugmentation, phytoremediation in soil) and ex situ biological treatments of polluted soils (biopiles and soil cultivation).

Keywords: soil, bioremediation, technology, biological, treatment

1. Introduction

Bioremediation is a modern pollutant treatment technology that uses biological factors (microorganisms) to transform certain chemicals into less harmful/hazardous final forms, ideally CO_2 and H_2O, which are non-toxic and are released into the environment without altering substantially the balance of ecosystems. Bioremediation is based on the ability of some chemical compounds to be biodegraded.

The concept of biodegradation is accepted as a summation of the decomposition processes of some natural or synthetic constituents, through the activation of some strains of specialized microorganisms resulting in useful or acceptable final products from the point of view of environmental impact [1–3].

In the last decades, the term bioremediation is used in a more specific way, which is reflected by the two specific definitions:

- The use of living organisms to degrade environmental pollutants, to prevent pollution or in the waste treatment process;

- Application of biological treatments for cleaning, decontamination and degradation of dangerous substances [3–5].

Bioremediation can be applied "in situ" (on the area, the polluted substrate, on the place where a contamination occurred) or "ex situ" (in specially arranged systems/installations, where it is brought the polluted substrate to be treated by biological methods).

2. Bioventing

Through bioventilation, oxygen is introduced into unsaturated contaminated soils through a forced air circulation (extraction or air injection) to increase the oxygen concentration and to stimulate biodegradation (**Figure 1**).

Bioventing is a technology that stimulates the natural in situ biodegradation of any aerobically degradable compound by providing oxygen to microorganisms existing in the soil. Compared to suction vapor extraction, bioventilation uses weak air flows to provide just enough oxygen to support microbial activity. Oxygen is most often applied by direct injection into residual soil pollutants. In addition to the degradation of adsorbed fuel residues, the volatile components are biodegraded as the vapor slowly circulates through the biologically active soil. Applicability of bioventilation is a medium and long term technology.

Cleaning can take from a few months to a few years. Bioremediation techniques through bioventing have been used successfully to remediate soils contaminated with petroleum hydrocarbons, non-chlorinated solvents, certain pesticides, wood preservatives and other organic chemicals.

Bioremediation can be used to change the valence state of inorganic substances and to cause adsorption, assimilation, accumulation and concentration of inorganic substances in micro- or macro-organisms. These bioremediation techniques are promising from the perspective of stabilizing or removing inorganic substances from

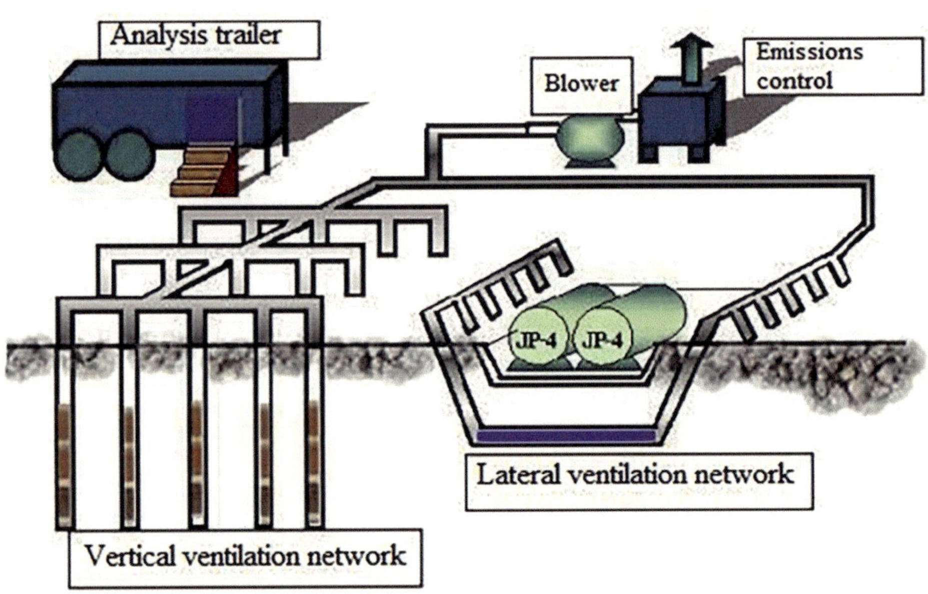

Figure 1.
Installation of bioventing [1] which includes: 1- analysis trailer, 2- blower, 3- emissions control, 4- vertical ventilation network and 5- lateral ventilation network.

the soil, while biodegradation cannot degrade inorganic pollutants. Limitations. Among the factors that can limit the field of application and the efficiency of the bioventilation process are: the mass of water at a few decimeters of the surface, saturated soil lenses or soils with low permeability reduce the efficiency of bioventilation. Vapors can collect in basins within the radius of influence of the air injection probes. This problem can be eliminated by vacuuming the air near the structure.

A low degree of moisture in the soil can limit biodegradation and the efficiency of bioventilation. It is necessary to monitor residual gases at the soil surface. Aerobic biodegradation of certain chlorinated compounds cannot be efficient unless there is a co-metabolite or an aerobic cycle.

Cold temperatures can slow down the cure, although successful cures have also been achieved in extremely cold environments. The condition is that a successful bioventilation is based on the fulfillment of two basic criteria. First, air must enter the soil in sufficient quantity to maintain aerobic conditions; and secondly, the microorganisms that naturally degrade hydrocarbons must be present in sufficiently high concentrations to achieve adequate biodegradation percentages.

The initial tests aim to determine both soil permeability and in situ respiration percentages. The grain size of the soil and its humidity have an important influence on the permeability of soil gases. The greatest restriction in terms of air permeability is represented by excessive soil moisture. The combination of high water tables, high humidity, and fine-grained soils prevented successful bioventing in certain locations. Among the soil properties that have an impact on microbial activity are pH, moisture and basic nutrients (nitrogen and phosphorus) and temperature. It has been calculated that the optimum soil pH for microbial activity falls between six and eight. The optimum humidity is established for each soil separately.

Too much humidity can reduce air permeability and decrease oxygen transfer capacity. Too low humidity will inhibit microbial activity. Several biovent tests have indicated biodegradation rates with moisture levels between 2% and 5% by weight. And yet, in extremely arid climates, it is possible to increase the rate of biodegradation through irrigation or humidification through injected air.

Pollutants degrade faster through bioventilation during the summer, but remediation can also occur at temperatures of 0°C. Hydrocarbon biodegradation rates are always estimated based on percentages of oxygen utilization, assuming that oxygen loss is due to microbial mineralization of hydrocarbons.

The main cost elements are the following: surface area; the number of installed injection/extraction wells matters. The number of wells (and related costs) increase proportionally with the area. Soil type; soil types with sand and gravel content reduce costs due to the smaller number of injection/extraction wells that need to be installed. Indicative prices for bioventilation fall between 25 and 200 Euros per cubic meter of soil. Costs can be influenced by the type of soil and its chemical properties, the type and amount of amendments used, and the type and extent of contamination.

3. Bioaugmentation

Bioaugmentation is a process in which local or inoculated microorganisms (such as fungi, bacteria and other microbes) degrade (metabolize) organic pollutants in soil and/or groundwater and neutralize their harmful effect (**Figure 2**). The activity of microbes that occur naturally is stimulated by the aqueous solutions that circulate through contaminated soils and that increase the degree of in situ biological

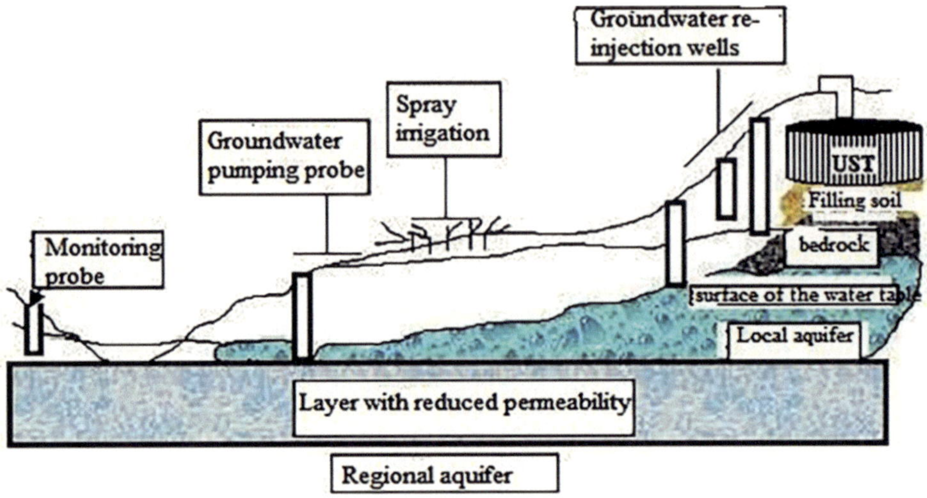

Figure 2.
Installation of bioaugmentation [1] which includes: 1- monitoring probe, 2- groundwater pumping probe, 3- spray irrigation, 4- groundwater re-jection wells.

degradation of organic pollutants or the immobilization of inorganic ones. Nutrients, oxygen and other amendments can be used to increase the bioremediation and desorption of pollutants from underground materials [1].

Bioaugumentation process is usually performed aerobically. In the presence of a sufficient amount of oxygen (aerobic conditions) and other nutrients, microorganisms will transform many organic pollutants into carbon dioxide, water and masses of microbial cells. Bioaugmentation of a soil normally involves the percolation or injection of groundwater or uncontaminated water mixed with nutrients and saturated with dissolved oxygen. Sometimes acclimatized microorganisms (bio-augmentation) and/or other sources of oxygen such as oxygenated water can be added. Sprinkler irrigation is regularly used for shallow contaminated soils, and injection wells for deep contaminated soils. Although in situ bioremediation has also proven successful in cold climates, low temperatures slow down the remediation process. Warm layers that cover the soil surface can be used for contaminated soils with low temperature to increase its temperature and the rate of degradation.

Bioaugmentation in the anaerobic process in the absence of oxygen (anaerobic conditions), organic pollutants will transform into methane, small amounts of carbon dioxide and tiny amounts of hydrogen. Under conditions of reduction with sulfates, sulfate is transformed into sulfide or elemental sulfur, and under conditions of reduction with nitrates it is finally produced into hydrogen sulfide.

Pollutants can sometimes degrade into intermediate or final products more or less dangerous than the original pollutant. For example, TCE (Trichlorethylene) can anaerobically biodegrade into vinyl chloride which is more persistent and toxic. To avoid such problems, most bioremediation projects are done in situ. Vinyl chloride can degrade further under aerobic conditions.

Bioaugmentation is a long-term technology that can take years to clean up a pollutant plume. Applicability in bioremediation techniques have been successfully used to remediate soils and sludge; remediation of groundwater polluted with petroleum hydrocarbons, solvents, wood preservatives and other organic chemical products.

Bioremediation is especially effective in remediating low-level residual contamination related to removal of the pollution source. The pollutant groups that were treated most often are PAHs, non-halogenated SVOCs (without PAHs), and BTEX (Benzene, Toluene, Ethylbenzene, Xylene (volatile organic compounds).

The types of contaminated sites treated most often were polluted by processes or through waste from wood preservation, oil refining, and recycling. Wood preservation involves the use of creosote, which contains a high concentration of PAHs and other non-halogenated SVOCs.

Similarly, oil refining and recycling processes frequently rely on BTEX. Given that the pollutant groups most often treated by bioremediation are SVOCs (PAHs and other non-halogenated SVOCs), treating them with volatility-based technologies such as SVE (soil vapor aspiration) could prove difficult. Bioremediation treatment does not frequently require heat treatment, involves few cost-effective elements such as nutrients, and does not normally generate residues requiring further treatment e or eliminations. Also, when done in situ, it does not require excavation of the contaminated environment. Compared to other technologies, bioremediation is advantageous in terms of price in treating non-halogenated SVOCs. Although bioremediation cannot degrade inorganic pollutants, it can instead be used to change the valence of inorganic substances and cause adsorption, immobilization in soil particles, precipitation, assimilation, accumulation and concentration of inorganic substances in micro- and macro-organisms.

These techniques show promise for stabilizing or removing inorganic substances from soil. Among the factors that prevent the applicability and efficiency of the process are: Cleanup objectives cannot be achieved if the soil base mass prevents pollutant-microorganism contact.

The circulation of aqueous solutions through the soil can increase the mobility of pollutants and may require treatment of the underground water table. Preferential colonization with microbes can cause clogging of nutrient and water injection wells. Many of the above factors can be controlled by paying attention to good technical practices.

The duration of the treatment can be from 6 months to 5 years and depends on factors specific to the respective site.

If bioaugmentation can achieve the cleanup goal in a compatible time frame, it can significantly reduce costs without excavation and transportation. Also, both contaminated groundwater and soil can be treated simultaneously, which is another cost advantage. In situ processes generally require longer periods of time, there is no certainty regarding the uniformity of treatments due to the inherent variability in the soil, the characteristics of the aquifer and the difficulties related to the monitoring process.

The remedial procedures can sometimes last for years depending especially on the degradation rates of the specific pollutants, the characteristics of the site and the climate. It could take less than a year to clean up certain pollutants, while higher molecular weight compounds take longer to degrade. Indicative prices for bioaugmentation fall between 25 and 220 Euros per cubic meter of soil. Costs can be influenced by the type of soil and its chemical properties, the type and amount of amendments used, the type and size of contamination [1].

4. Phyitoremediation

Phytoremediation is a process of using plants to remove, transfer, stabilize and destroy pollutants from soil and sediments. Pollutants can be organic or inorganic (**Figure 3**).

Figure 3.
Phytoremediation in soil organic and inorganic [1].

Phytoremediation is a process of using plants to remove, transfer, stabilize and destroy pollutants from soil and sediments. Phytoremediation mechanisms include advanced rhizosphere biodegradation, phytoaccumulation, phytodegradation and phytostabilization [1–3].

In addition, poplars can draw large amounts of water (compared to other plant species) when it passes through the soil or directly from the aquifer. This means absorbing a large amount of dissolved pollutants from contaminated environments and reducing the amount of pollutants dispersed through or out of the soil or aquifer. Phytoaccumulation represents the assimilation of pollutants by plant roots and their movement/accumulation (phytoextraction) in the trunk and leaves. Phytodegradation is the metabolism of pollutants in plant tissues.

Plants produce enzymes such as dehalogenase and oxygenase that help catalyze degradation. Investigations will determine whether aromatic and chlorinated compounds respond to phytodegradation. Phytostabilization is a phenomenon of the plant's production of chemical compounds that serve to immobilize pollutants when the roots come into contact with the soil.

Phytoremediation can be applied to remediate metals, pesticides, solvents, crude oil, PAHs and landfill leachate. Some plant species have the ability to store metals in their roots. These plants can be transplanted to contaminated sites to filter metals from wastewater. When the roots become loaded with metal pollutants, these plants can be removed.

Plants that accumulate large amounts can remove or store significant amounts of metal pollutants. Currently, trees are being tested to determine their ability to remove organic pollutants from groundwater, translocation and transpiration and their possible metabolism into CO_2 or plant tissues. Soil phytoremediation can be limited by: the depth of the treatment zone is determined by the plants used in the phytoremediation. In most cases, this method can be used on shallow soils. High concentrations of hazardous substances can be toxic to plants. It has the same mass transfer limits as other biotreatments.

Sometimes it can only be done in certain seasons, depending on the locations. It can transfer pollutants between environments, such as from soil to air.

It is not effective for strongly absorbed pollutants (such as polychlorinated biphenyl PCBs) and poorly absorbed ones. The toxicity and bioavailability of degradation products are not always known. The products can be mobilized in groundwater or bioaccumulated in animals. Detailed information is needed to determine the soil types used in phytoremediation projects. Flow, oxygen-reducing concentrations, root growth and their structure affect plant growth and must be taken into account when implementing phytoremediation. Performance data. Several phytoremediation demonstrations are currently being done (e.g., oak species were planted in the middle of a TCE pollutant plume to assess TCE transpiration and TCE transformation rates). Evaporation-transpiration rates were measured equally.

This is a long-term test of the ability of trees to control the circulation of underground water. Important cost elements Degree of effort; the contaminated area influences the costs the most. Sampling density; essential cost element for determining sample costs; can be directed by regulatory requirements. Phytoremediation is a long-term remedial process. Costs are mainly caused by procurement of plants, investments related to trees with subsequent testing and sampling costs. Costs can vary from 10 Euros for a lightly contaminated site to 150 Euros for a difficult site, per cubic meter treated [1].

5. Biopiles

The excavated soils are mixed with amendments and placed in enclosures on the surface. It is a composting process with aerated static piles in which the compost is raised in piles and aerated with blowers or vacuum pumps (**Figure 4**).

Suppliers have developed proprietary nutrient and additive formulas, as well as methods of incorporating the formulas into the soil to promote biodegradation. Formulas are usually modified from soil to soil. Soil mounds and piles usually have an air distribution system buried underground to allow air to pass through the soil

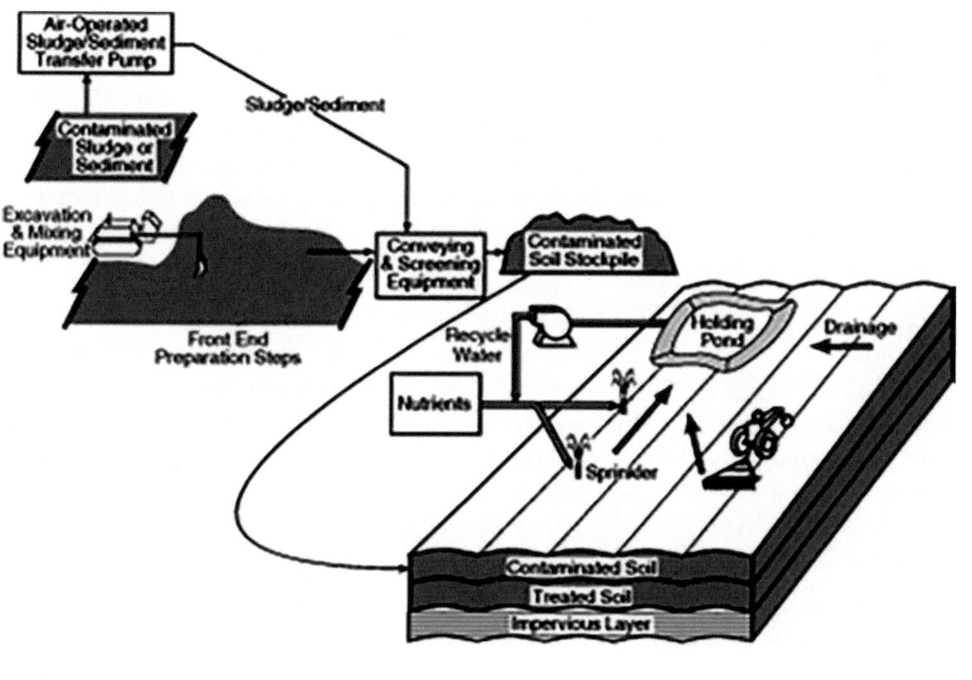

Figure 4.
Typical biopile system for solid phase bioremediation [1].

either through vacuum or positive pressure. In this case, the mounds of soil can have a maximum height of 2–3 meters.

These mounds can be covered with plastic to control scattering, evaporation and volatilization and to stimulate solar heating. If there are VOCs in the soil that will evaporate into the air, the air that is emitted from the soil can be treated to remove or destroy the VOCs before entering the atmosphere. Biopile is a short-term technology that can last from a few weeks to a few months. Treatment options include static processes such as treatment bed preparation, biotreatment cells, soil mounds, and composting.

Biopile treatment has been applied to non-halogenated VOCs, hydrocarbons from fuels, halogenated VOCs, SVOCs, pesticides can also be treated, but the efficiency of the process will vary and may only be applicable to a few compounds within these pollutant groups.

Among the factors that can limit the applicability and efficiency of the process are:

- Excavation of contaminated soils is necessary.
- Treatment grade testing should be performed to determine pollutant biodegradability, adequate oxygenation, and nutrient loading rates.
- Solid phase processes are likely effective on halogenated components and may prove ineffective in degrading explosives transformation products.
- Piles of similar size take longer to clean than sludge phase processes.
- Static treatment processes may result in less uniform treatment than processes involving periodic mixing. The first steps in preparing a well-argued project for the biotreatment of contaminated soils include: site characteristics and soil samples and characteristics.

Biopile treatment has been demonstrated on several fuel-polluted sites. Costs depend on the pollutants, the procedure to be applied, the need for additional or post-treatment and the need for air control equipment. Biopiles are quite simple and require little personnel for maintenance and handling. Typical indicative costs with one coat and a prepared liner are between 35 and 100 Euro per cubic meter [1].

6. Composting

Contaminated soil is excavated and mixed with fill materials and organic amendments such as wood scraps, hay, natural fertilizers and plant waste (e.g., potatoes). Selecting the right amendments ensures sufficient porosity and provides a balance between carbon and nitrogen to promote thermophilic microbial activity (**Figure 5**).

Composting is a controlled biological process through which organic pollutants (for example PAH) are transformed by microorganisms (in aerobic and anaerobic conditions) into harmless, stabilized by-products. Normally, thermoelectric conditions (54 to 65°C) must be maintained to properly fertilize soil contaminated with hazardous organic pollutants. The high temperatures result from the heat produced by microorganisms during the degradation of organic matter in the waste. In most cases, this is achieved by using local microorganisms. Soils are excavated and mixed

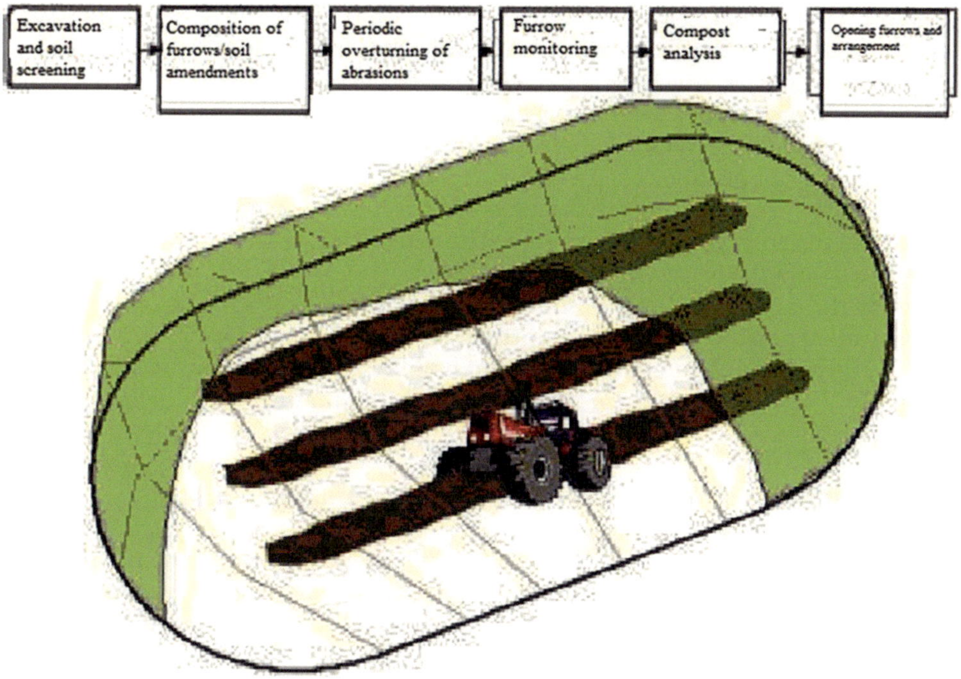

Figure 5.
Coposting scheme [1] which includes: Excavation and soil screening, composition of furrows/soil amendaments, periodic overturning of abrasions, furrow monitoring, compost analysis and opening furrows arrangement.

with filling materials and organic amendments such as sawdust, animal and vegetable waste, in order to increase the porosity of the mixture that will be decomposed. Maximum degradation efficiency is achieved by maintaining oxygenation (such as daily furrow turning), irrigation if necessary, and careful monitoring of soil moisture and temperature [1].

There are three types of processes used in composting: fertilization of static aerated mounds (the compost is raised in piles and aerated with blowers or suction pumps), by composting in mechanically stirred vessels (the compost is placed in the reactor vessel where it is mixed and aerated), and furrow composting (compost is placed in long mounds known as furrows that are periodically mixed with mobile equipment). Furrow fertilization is usually considered the most cost-effective composting option. At the same time, it can also present thousands of transient eruptions.

If VOCs or SVOCs are present in the soil, off-gas control may be required. The composting process can be applied to soils polluted with biodegradable organic compounds. Aerobic, thermophilic composting can be used for PAH-contaminated soils. Any material or equipment used in composting is available on the market.

Factors that can limit the applicability and efficiency of the process include: A large space is required. Excavation of contaminated soils that may cause uncontrolled VOC emissions is required. Composting leads to an increase in volume of the material due to the addition of amendments. Although metal levels can be reduced by dilution, heavy metals cannot be treated by this method. High concentrations of heavy metals can be toxic to microorganisms.

Among the specific data needed to evaluate the composting process are pollutant concentration, excavation requirements, availability and cost of amendments required for the compost mixture, space required for treatment, soil type and pollutant response to composting. Furrow composting has been demonstrated as an effective technology for treating soils contaminated with explosive substances.

The cost of providing a treatment base with the collection of polluted seepage water is included. The main cost elements are: The type of pollutant is the key element in determining composting costs. Soil type/total organic content (TOC); soils with higher density (generally fine sands and gravel) cost less to compost, while soils with higher TOC would require more expense. The density influences the mass of the soil to be treated, and the percentage of TOC indicates the level of contamination. Costs depend on the pollutants, the procedure to be applied, the need for additional or post-treatment and the need for air control equipment. Biopiles are quite simple and require little personnel for maintenance and handling. Typical indicative costs with one layer and a prepared liner are between 35 and 100 Euro per cubic meter [1–3].

7. Conclusions

Bioventilation has become a common technology, with most of the technical components already available. Bioventing is receiving a lot of remedial attention from the consulting community, especially regarding the use of this technology in conjunction with soil vapor extraction (SVE). As in the case of all biological technologies, the time required to remediate a site through bioventilation depends to a large extent on the characteristics of the soil and the chemical properties of the contaminated environment. With bioaugmentation there is a risk of increasing the mobility of pollutants and their infiltration into the water table. Bioaugmentation has been selected for corrective and emergency actions on a large number of contaminated sites. In general, petroleum hydrocarbons can be immediately bioremedied at relatively low cost by stimulating local microorganisms with nutrients.

Author details

Ioana Stanciu
Faculty of Chemistry, Department of Physical Chemistry, University of Bucharest, Bucharest, Romania

*Address all correspondence to: istanciu75@yahoo.com

References

[1] Guérin V, Menger P. Depoluarea solului şi a apelor subterane poluate cu compuşi organici. BRGM, Franța; IHOBE, Țara Bascilor; 2010. Available from: http://www.timisoara-twinningenvironment.net/media/dms/File/depollutionROweb.pdf

[2] Dar A, Naseer A. Recent applications of bioremediation and its impact. In: Hazardous Waste Management. IntechOpen; 2022. [Internet] DOI: 10.5772/intechopen.104959

[3] Tyagi B, Kumar N. Bioremediation: Principles and applications in environmental management. Bioremediation for Environmental Sustainability. Jan 2021. pp. 3-28. DOI: 10.1016/B978-0-12-820524-2.00001-8

[4] Azubuike CC et al. Bioremediation techniques–classification based on site of application: Principles, advantages, limitations and prospects. World Journal of Microbiology and Biotechnology. 2016;**32**:180

[5] Gupta C, Prakash D. Novel Bioremediation Methods in Waste Management: Novel Bioremediation Methods. Waste Management. Jan 2020. pp. 1627-1643. DOI: 10.4018/978-1-7998-1210-4.ch075

Chapter 4

Standard Analytical Techniques and *de novo* Proposals for Successfull Soil Biodegradation Process Proposals

Juan Cabral-Miramontes, Pamela Dorantes-Alvarado and Elva Teresa Aréchiga-Carvajal

Abstract

The contamination of water, air, and soil represent a serious problem worldwide. Therefore, it is a priority to reduce the levels of cytotoxic in the environment caused by human activities that generate chronic degenerative diseases. For example, soil contamination caused by oil and derivatives removed with biotechnological products based on biological systems of microorganisms with physiological and molecular mechanisms that allow them to carry out effective bioremediation processes, reducing the concentration of polluting hydrocarbons. The main obstacle is validating the biodegradation efficiency of chemical compounds by bacterial consortia; therefore, it is vital to adapt or develop analytical strategies to verify heavy-end reduction for each type of biological system used in remediation. This chapter describes the techniques and their adaptations for oil degradation and their derivatives promoted by microorganisms. As the limits of the methods vary within the parameters determined by international norms and laws, we compare conventional and new-generation proposals to adjust to probe biotechnological products based on consortia of biodiverse microorganisms that significantly degrade petroleum fractions.

Keywords: microorganisms, bioremediation analytical, soluble lipase, soil biodegradation, analytical techniques

1. Introduction

Anthropogenic activities for the development of the economy, such as the extraction, storage, transportation, and general use of fossil fuels and their derivatives, are the primary sources of environmental pollution. Affected natural sites are bodies of water and soils close to anthropogenic activities [1]. Industrial growth causes disturbances and the loss of species of flora and fauna, that is, pollution causes the extinction of species that have an essential value in food chains for the conservation of biodiversity, an issue that is no less critical at the global level world [2].

Environmental legislation for the development of a sustainable economy is the challenge and problem faced by politicians and scientists from developed countries, such as the European Union and the United States of America, and developing countries, such as Mexico, these regions do not reduce articles derived from petroleum due to the increase in its population [3].

Microorganisms have a vital ecological function in biomes, and in addition, they represent a source of mega diversity in the environmental niches in which they live. Microbes provide humans with privileged information to determine biotechnological uses due to their physiological characteristics derived from their adaptation and molecular evolution, which allows them to survive in hostile environments where the bioremediation of sites contaminated by petroleum derivatives is to be used [4].

Microbes eliminate contaminants such as pesticides, herbicides, heavy metals, hydrocarbons, and plastics. The first step is to evaluate the type of contaminant, concentration, and *in situ* conditions to generate successful bioremediation. Second, to determine the bioavailable nutrients the microorganisms could feed, it is essential to identify the exact values of abiotic factors such as pH, temperature, UV rays, and salinity at the site of remediation by microbes [5]. With this information, we can designate the strains or consortia strains of bacteria and fungi to be used to generate efficient biological processes to eliminate environmental contaminants in affected soils [6].

Associations of bacteria use specific enzymes such as lipases to hydrolyze hydrocarbons into fatty acids and glycerol [7]. The main structural features are the consensus amino acid sequence of Gly-X-Ser-X-Asp and the chaperone protein Lif (lipase-specific foldase) in charge of establishing disulfide bonds [8]. Lipases have an alternate α/β hydrolase domain, a catalytic triad of Ser, His, and Asp residues, and high activity in the presence of tributyrin and oils [9]. Function of lipases depends on factors such as temperature, pH, aeration, and the presence of ions that can improve the activity of splitting hydrocarbons and their derivatives [10]. Therefore, locating water-soluble lipases, or lipases immobilized by physical or chemical methods, in bacterial genomes to increase the efficacy of bioremediation in contaminated environments is a promising alternative for contemporary problems.

In this chapter, we describe the biological remediation systems, the size and solubility of the molecules when they are degraded by bacteria, the reported genes that are involved in the bio generation of enzymes used by microorganisms for degradation, and the most precise analytical methods for the different cases of bioremediation, in addition to the international laws and regulations for the regulation of the effective use of microorganisms.

2. Bacterial removing metabolism for oil and hydrocarbons degradation

Hydrocarbons and their derivatives strongly affect the environment. The most abundant forms are crude oil, gasoline, kerosene, diesel, and oils, whose presence in the soil due to spills reduces the viability of any form of life in growth. In addition, the presence of hydrocarbons can be distinguished by their specific odor from kilometers away, generating a warning zone for the population near the site. However, by not taking the necessary precautions in the contaminated area, it becomes a problem for public health because it causes damage to human health, generating diseases of carcinogenic origin that are heritable between generations of people [11]. Microorganisms represent an alternative with an advantage over plants and other bioremediation processes because they successfully survive contaminated sites. They modify their environment and use hydrocarbons as a source of nutrients. Among the most recognized species of microorganisms for their tolerance and elimination of hydrocarbons are *Pseudomonas* and *Aspergillus* [12].

The increased demand for contaminated surfaces leads to the search for niche providers of microbes with genomic and metabolic evolution with the ability to unfold and eliminate hydrocarbons using these molecules as a carbon source. The investigation of these microorganisms' biodegradation capacity and metabolic plasticity generates information on the optimal conditions and crucial factors associated with their success in contaminated sites, such as temperature, oxygen, pH, and nutrients. When optimal conditions are present, degradation is older [13].

The metabolic interactions of bacteria-bacteria or bacteria-fungi consortia represent a green strategy to eliminate the highest percentage of contaminants under aerobic and anaerobic conditions since some microbes show adaptation to a specific hydrocarbon derivative as a carbon source for cell growth. *Acinetobacter* sp., *Brevibacterium*, *pseudomonas* sp., *Aspergillus* sp., and *Candida* sp. can degrade oil compositions such as aliphatic hydrocarbons are saturated or unsaturated. Aromatics are ringed hydrocarbon molecules divided into (a) monocyclic aromatic hydrocarbons (MAH), BTEX (benzene, toluene, ethylbenzene, and xylenes), and (b) polycyclic aromatic hydrocarbons (PAHs) (**Figure 1**), which are degraded by microorganisms such as *Bacillus* sp., *Halomonas* sp., and *Rhodococcus* sp. Finally, bacteria belonging to families such as *Vibrionaceae* and *Enterobacteriaceae* resins degrading. The metabolic processes of bioremediation by microorganisms occur in the decomposition of molecules to generate easily eliminated by-products. The two main hydrocarbon degradation pathways are aerobic and anaerobic [14, 15].

Aerobic degradation represents various metabolic routes where most reactions are oxidations, causing hydrocarbons to break down into smaller molecules. Molecules of up to 12 carbons are usually more challenging to eliminate as a source of nutrients for microorganisms [16]. Aerobic metabolism of hydrocarbons begins with the oxidation of the contaminant and oxygen incorporation, oxidation participation, and

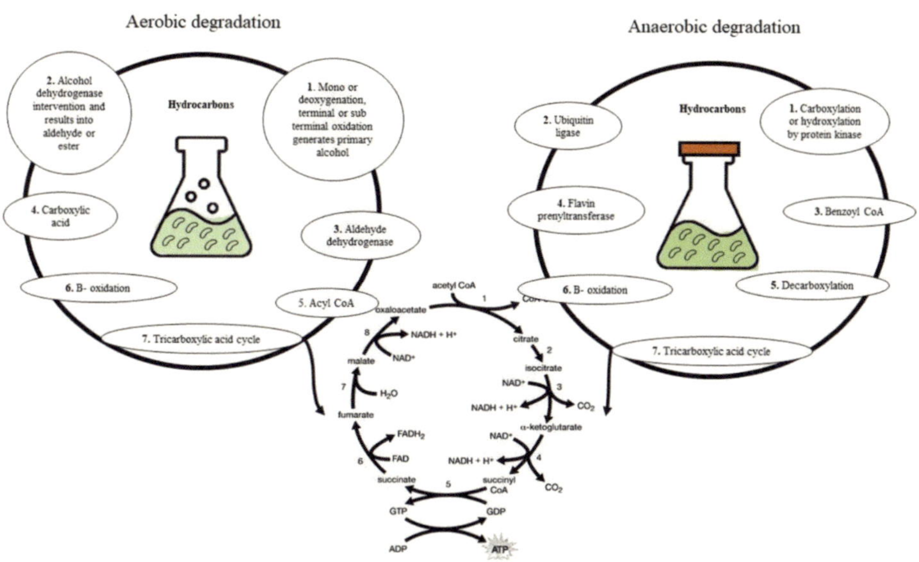

Figure 1.
Most remarkable reactions for degradation of hydrocarbons for aerobic and anaerobic degradation [14, 15].

B-oxidation. Diversity of available hydrocarbons implies locating efficient and specific metabolic routes located in the bacterial genome that allows them to degrade hydrocarbons such as octane, decane, alkane, pristane, eicosane, phenol, benzoate, generate, benzene, toluene, and xylene. The second part is the segmentation of a specific type of hydrocarbons. This process eliminates it *via* alkane molecule reduction; later, they are catalyzed by the enzyme alkane hydrolase (AH) and obtain fatty acids. Finally, the acyl CoA becomes susceptible to degradation by the acyl-CoA synthetase intervention. This way, the bioremediation niche receives small molecules for easy degradation as a final product, commonly used in some energetic metabolic processes [17].

Contrary to the previous case, oil anaerobic degradation presents the same speed as aerobic degradation. Metabolism, in this case, is led by the oxidation of pollutants to phenols, and organic acids are transformed into long-chain fatty acids, finally ending in CH_4 and CO_2. In addition, we must consider that different metabolic pathways must coincide in microbial genomes processes. Some ions are also considered, including nitrate, ferrous iron, manganese, and sulfate. For alkanes and cycloalkanes, the reaction starts with the exchange of fumarate at the subterminal methylene carbons and a straight-chain hydrocarbon into a branched compound [18, 19].

3. Characteristics of *n*-alkanes for bacterial uptake

The mechanisms of bacteria to capture alkanes depend on the species and the ecological niche where it is located (**Figure 2**). It is mediated by the alkane's molecular weight that enters the soil through spills caused by human carelessness [20]. The water solubility of *n*-alkanes changes due to their physicochemical characteristics, and as their molecular weight increases, the water solubility decreases. The viscosity of alkanes represents a challenge for absorption by microorganisms to transform inert alkanes into low molecular weight substances that are oxidized with enzymes produced by microbial populations [16].

Soil characteristics also determine changes in the conditions and accessibility of *n*-alkanes, hindering accessibility by microorganisms and their biodegradation.

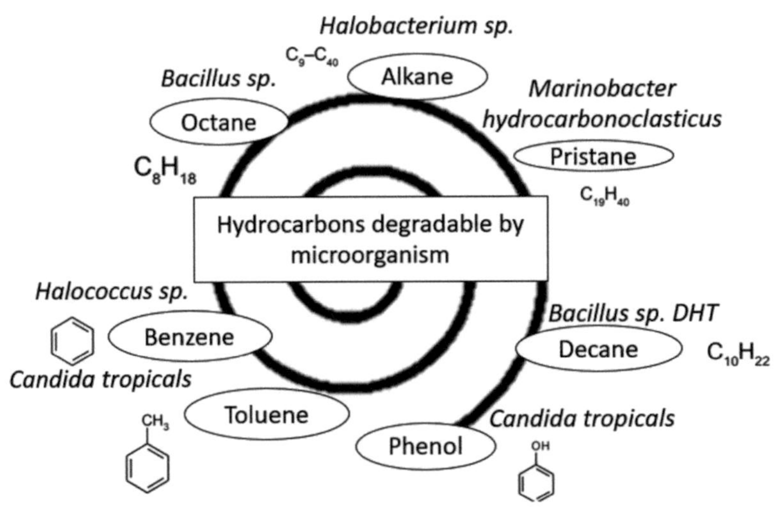

Figure 2.
Some of the hydrocarbons and the microorganisms that are capable of degradation.

The way of degradation through efficient biological processes is to know the weight of the molecules derived from petroleum present in the soil and the time they have remained in the site, which is due to the solubility in water. The commercially used alkane isoforms of which we can find in a spill reported by Rojo [21].

Microorganisms use enzymes such as peroxidases, laccases, monooxygenases, ligninolytic, and hydrolases to break down molecules of considerable molecular weight into simpler ones and use them as a carbon source for their nutrition [22]. Biological hydrocarbon degradation processes are promising techniques for the bioremediation of environmental contamination caused by hydrocarbons. However, it is necessary to consider the environmental conditions to which the treatments are exposed, that is, the degraded alkane and the physical-chemical conditions of the soil, to choose with certainty the necessary processes of microorganisms that allow the degradation of petroleum derivatives in various polluted environments.

4. Mechanisms involved in the degradation of polycyclic aromatic hydrocarbons

Microorganisms degrade hydrocarbons through the catalysis of intracellular enzymes, reducing molecules from high to low molecular weight [23]. The first step is of great importance; the microorganisms detect the contaminant and initiate the secretion of tens of active molecules to carry out the emulsification, such as the rhamnolipids bio-produced by *Pseudomonas aeruginosa*. The microorganisms then absorb the emulsified contaminants through the cell wall; subsequently, they are endocytosed to generate active or passive intracellular transport; finally, they undergo an enzymatic reaction with specific enzymes for each molecule to complete the degradation process [24].

Aerobic degradation is centered on three modes of oxidation: (i) terminal oxidation, in which it produces primary alcohol that later, through another reaction, becomes an aldehyde, carried out by aldehyde dehydrogenase, it becomes a fatty acid, on the other hand, and in a longer path [25]; (ii) subterminal oxidation generates secondary alcohol to methyl ketone, and later form an acetyl-ester through monooxygenation, likewise with the function of an esterase enzyme, it can divide a molecule into two parts, creating primary alcohol plus an acetate, to finally generate a fatty acid [26]; (iii) biterminal oxidation is a more straightforward enzymatic process where, through a pyruvate carboxylase enzyme, it produces carboxylic acid as a final product [27]. The end products of the three pathways enter beta-oxidation, as degradation progresses, and simpler molecules of low molecular weight are produced, which are used intracellularly by microorganisms as an energy source [28].

4.1 Bacterial lipase

Enzymatic activity with lipases is under diverse conditions such as temperature and good oxygenation. More importantly, lipids are a carbon source, and some trace elements enhance lipase production. Lipase from different bacteria is vital in the industry, *Bacillus pumilus* in the food and detergent industry [29], *P. aeruginosa* in solid waste treatment [30], and *Staphylococcus hemolytius* and epidermidis for the food industry [31] strains are supplemented with substrates, such as long-chain triacylglycerols, triolein, corn oil, or fish oil, to maximize their hydrolytic activity [32]. Time and carbon sources are essential for their growth. In long incubation times, an increase in lipase production has been found, in addition to obtaining different molecules or isoforms [33]. The central carbon chain is reduced at the ends when a lipase reaction generates glycerol and/or a shorter fatty acid. Consequently, with each bacterial strain, conditions may vary. *Pseudomonas fluorescens* performs optimally at temperatures of 70°C [8].

The influence of metal ions and trace elements shows an active presence of lipase [34, 35]. Some of the most studied metal ions are sodium and calcium, which have been shown to significantly influence lipase production and long-chain fatty acid hydrolysis; furthermore, metal ions such as calcium stabilize the enzyme structure (**Figure 3**).

Lipase activity applied to bioremediation effectively reduces contaminants such as oil, especially hydrocarbons and reused oils, on surfaces encountered. There are two strategies for the use of lipase, first is the enzymatic enrichment directly on a soluble substrate and second is the insolubility of the oil and its derivatives in the use of the enzyme in immobilized form, where we know several options to immobilize lipase, for example, under soluble substrates, such as the sectional process where it involves the addition of water and the presence of alcohol so that the enzyme catalyzes both reactions at the same time: hydrolysis and esterification [33]. Soluble lipase influences the replies above, where it has high catalytic activity in the first hours of contact. The movement loses during the first time (4–5 hours) due to the denaturing action [8].

Immobilized lipases show multiple benefits, such as excellent thermal and ionic stability and efficient enzyme control in specific reactions [34]. Enzymes bind to substrates for biological adsorption processes during a natural removal process. For its success, the union achieves cationic and anionic exchange resins activated by carbon molecules. Also, a low-cost silica gel method is generated, is diffusion controlled, and occurs naturally under normal bacterial conditions [35, 37]. The encapsulation increases the contact surface between the enzymes, the substrate, and

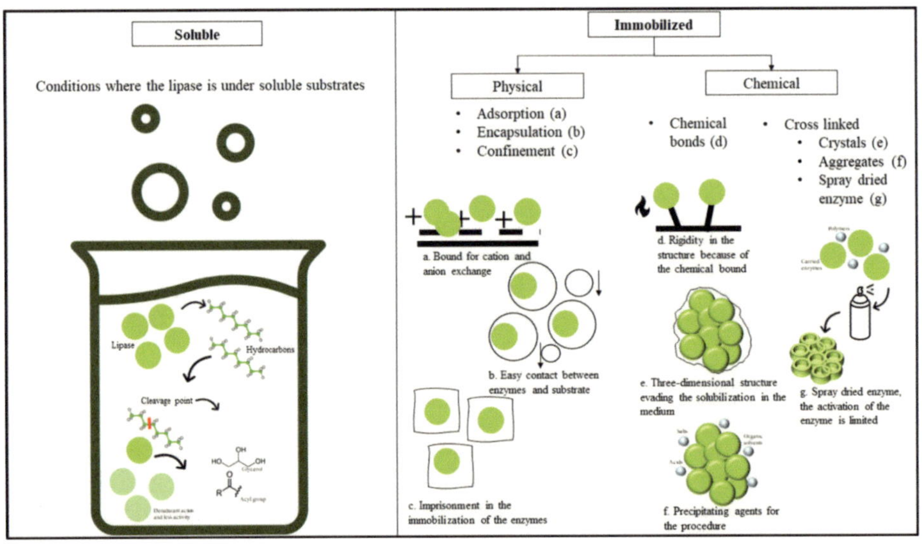

Figure 3.
Review of the uses of lipase in soluble and immobilized conditions. Soluble conditions and the interaction within the hydrocarbons [8, 36]. Immobilized strategies, physicals: (a) absorption [35, 37–40], (b) bncapsulation [37, 40], (c) confinement [41, 42], (d) chemical: chemical bonds [42], cross-linked with (e) crystals, (f) aggregates, and (g) spray dried enzyme [43–46]. Reference articles for a drawing of the picture are listed in each part.

the mechanical stability. The disadvantages of these methods are the deactivation of the enzyme during the abrasion of the support material and the small load capacity [41]. Confinement is the latter method; the physical immobilization of enzyme bonds was part of a polymerized reaction mixture when compounds were formed. Then, the porous matrix includes the site of the biocatalyst to be immobilized and surrounds the enzyme, confining it in its structure and the substrate [47].

Chemical methods infer chemical bonds from reactions such as glutamic acid, lysine, cysteine, and aspartic acid residues between vehicle ingredients. The technique increases stability and ensures rigidity in the structure, avoiding being affected by denaturing agents such as organic solvents and heat. (i) The solution adds the crystallized enzyme, and stabilization arises, forming a three-dimensional solid structure [42, 43]. (ii) The cross-linked enzyme adds a precipitating agent such as salts, acids, and organic solvents, which use protein precipitation. (iii) A spray-dried cross-linked enzyme, a polymer, and a solution containing the enzyme into a spray dryer, enzyme deactivation occurs during spray drying of the application [44–46].

4.2 Genes that code for the biosynthesis of enzymes involved in the degradation of PAHs

The central system of alkane degradation by bioremediation of polluted environments is AH, *n*-alkanes in alcohols are hydrolyzed, and then fatty acids are oxidized to enter beta-oxidation [48]. In bioremediation by microorganisms, the monooxygenase (AlkB) family of enzymes [49] has essential components such as rubredoxin and rubredoxin reductase, which interact with electron transfer to carry out hydroxylation on molecules of 10 to 16 carbon atoms [50]. The CYP153 family of cytochrome proteins degrades short and medium-chain *n*-alkanes (C5–C10) [51]. Studies by Nie [52] and collaborators in 2014 reported a total of 3,979 bacterial genomes analyzed, locating 458 genes that code for AlkB. Structural analysis of the genes confirms the presence of families of orthologs containing an AlkB domain, also a family including an AlkB domain + a Rubredoxin (Rd) domain. In addition, they determined a family classified as a protein with three structural components: ferredoxin (Fer) + ferredoxin reductase (FNR) + AlkB; likewise, in the analyzed data, they found 130 genes for CYP153, having only two variants in its structure, a family that had an N-terminal domain of cytochrome P450 (CypX) and a family with CypX + FNR + Fer domain. Finally, they detected 73 and 32 genomes with multiple copies of the AlkB and CYp153 families. Variations in the conserved domains of genes that code for degrading enzymes are the success in the biological processes of degradation of hydrocarbons and their derivatives present in the soil [52].

Ji [53] and collaborators mention that terminal or subterminal oxidation promotes the aerobic catabolism of *n*-alkanes generated by the bacterium pseudomonas aeruginosa. In these reactions, AH enzymes of the families alkB1, alkb2, rubredoxin rubA1, rubA2, and rubredoxin reductase (rub) are the ones that carry out the degradation. Grady et al. [54] analyzed these enzymes in two strains: the *P. aeruginosa* strain ATCC 33988 and an environmental isolate (PAO1, which comes from storage tanks). A comparison of gene sequences of enzymes involved in the degradation of *n*-alkanes shows 99% identity. At the same time, they compare the expression levels of the two genes as mentioned above; they determined that transcription up-regulated in cultures with *n*-alkanes, which generates precise information that alkB1 and alkB2 catalyze the initial oxidation step in their degradation, that is, a successful downgrade. Physiological behavior shows an improvement in survival of strain ATCC 33988 compared to strain PAO1, generating approach

Instrumental analysis	Acronym	Description	Samples examined	Microorganisms	Reference
Gas chromatography	GC	It works to determine patterns of distribution of hydrocarbons in the samples, and its success depends on adequate preparation of the soil sample, where optimal solvents are used. The identification of compounds achieves by comparison with predicted hydrocarbon patterns.	Paraffins, phenol, anthracene, and pyrene	*Rhodococcus erythropolis*, *Rhodococcus cercidiphyllus*, *Arthrobacter sulfuroso*	[55]
Gas chromatography-mass spectrometry	GC-MC	It is a highly informative analytical technique that provides accurate information on organic compounds in small samples. The determination of hydrocarbons requires the presence of databases of mass spectra.	Soil contaminated with motor oil.	*Cronobacter sakazakii*	[56, 57]
Comprehensive two-dimensional gas chromatography	GCxGC	This technique obtains information in two columns, first, the column uses a non-polar solvent to separate according to a boiling point, and the eluting sample is sent to a second short (polar) column for separation by polarity.	Booth [58] has used GCxGC-TOFMS to reveal the complex hydrocarbon mixture from the contaminated mussels collected	N/A	[59]
Liquid chromatography and liquid chromatography-mass spectrometry	HPLC/ LC-MC	This analysis is selected when the sample has liquid properties; therefore, its polarity is not stable; that is, its application is minimal for the study of hydrocarbon biodegradation in soils, and its analytical principle involves the interaction of UV (absorption or fluorescence) with the compound to be analyzed.	Degradation of PAH and phenanthrene in soil samples	*Pseudomonas stutzeri*	[60, 61]
Infrared spectroscopy	FTIR	It is a technique used to determine the degradation rates of TPH, PAH, and alkanes present in the soil. Its foundation is to show the viscosity and reduction by degradation from the measurement of the interaction of infrared radiation with absorption material, and it is considered a nondestructive and straightforward test.	Oil-contaminated soils	*Clostridium* spp.	[62–64]
Magnetic resonance spectroscopy	NMR	A technique for identifying, determining, and characterizing metabolites secreted by the presence of hydrocarbons derived from petroleum, identified as 1-acenaphthenol, 1-acenaphtenone, acenaphthene-1,2-diol, and naphthalene 1,8-dicarboxyl acid.	Oil-contaminated soils	*Burkholderia cepacia*	[65, 66]

Table 1.
Analytical methods used in the determination of degradation by a microorganism.

information to discover candidate genes for growth enhancement with metadata (omics) tools. With these solid results, they report the expression of the aprX and aprA operon that shows activity in the presence of *n*-alkanes, both proteins are proteases secreted to degrade the *n*-alkanes present in the medium [54].

5. Analytical methods for the detection of HAPS degradation by microorganisms

The success of hydrocarbon bioremediation depends on the degradation carried out by microorganisms due to the evolutionary adaptation they show in extreme conditions. Therefore, it is essential to examine the compounds, which are still in the contaminated area, and control their degradation during biological removal processes. Analytical techniques (**Table 1**) represented by Imam et al. [67] are vital for measuring quantitative results from microorganisms.

Techniques to be used in biodegradation tests depend entirely on the availability of the samples obtained, the contaminant's origin, texture, the degrading capacity of the microorganisms, and their metabolic pathways. And finally, it also depend on budgets to carry out reliable analytical techniques.

6. Regulation of the use of microorganisms in Mexico, the USA, and Europe

The contaminants present in the water and the soil, as well as their remediation, have been an urgent problem to solve; many contaminants can be removed from the surface using different methods, such as physical, chemical, and biological. However, many of these tend to leave remnants that still affect soil and water rather than complete remediation; the regulation of these methods and the proposal of new ones as the use of microorganisms has increased over the years [68].

6.1 Mexico

In Mexico, associations regulate and administer all damage to soils and waters (**Table 2**). The results of remediation promote specific laws; the Ministry of Environment and Natural Resources (SEMARNAT) is an important institution for its wide application to different fields of natural science and conservation.

6.2 The United States of America

The United States of America designed an association regulating the pollution caused by soil and water. It has a manual with recommendations. This institute is the United States Environmental Protection Agency (EPA), and it reports updates every year, developments, and the latest changes in environmental laws (**Table 3**).

6.3 Europe

Europe is mainly concerned about the use of microorganisms and has a classification of their benefits, accepting all the strategies used for bioremediation. The organization for the management and distribution of information is the "Ministry

Norm	Description	Reference
NOM-138-SEMARNAT/SSA1-2012	Limits of hydrocarbons on soil, characterization, and aspects for their remediation.	[69]
NOM-035-SEMARNAT-1993	Methods for measurement of particle concentration suspended in the air and calibration of the equipment. The first submission for a proposal of treatment is made through this norm.	[70]
NOM-147-SEMARNAT/SSA1-2004	Set the criteria for the characterization and determination of remediation concentration in contaminated soil and the parameters of the contaminant present in the soil.	[71]
NOM-052-SEMARNAT-2005	Characteristics, identification procedures, classifications, and a list of harmful contaminants.	[72]
SEMARNAT Art. 134, 135, 138, 143	Investigation of the contaminant and the incidence of the event, recent characterization of the soil and contaminant, environmental risk, and the remediation treatment proposal.	[73]

Table 2.
Mexican regulation for the use of techniques where bioremediation is involved. Normative representative [69–73].

Norm	Description	Reference
EP A 542-F-21-028	Presents the best management practices of green remediation and classifications.	[74]
EPA 524-R-018	Introduction to *in situ* bioremediation of groundwater as a reference for the investigation.	[75]
EPA 542-f-16-001	Development and monitoring of contaminated water.	[76]

Table 3.
USA regulation for the use of techniques where bioremediation is involved. Normative representative [74–76].

Norm	Description	Reference
Law 22/2011 Art.24	Management of bio residual and the protection of the environment.	[77]
Law Real Decree 9/2005	Contaminated soils and establishment of conditions for making effective the protection of the environment.	[78]
Real Decree 664/1997	Defines the use of microorganisms and their classifications as well as the precautions.	[79]

Table 4.
Europe regulation for the use of techniques where bioremediation is involved. Normative representative [77, 78].

for Ecological Transition" (**Table 4**). Europe has a comprehensive website that shares different regulations, updates, and instruction manuals.

The Ministerium for the Ecological Transition and Demographic Challenge offers its website https://www.miteco.gob.es for knowledge and updates. Awareness of the environmental situation is an urgent problem and needs progress. Moreover, many countries reached advances in applied technologies and regulation of it.

The differences between Mexico, the USA, and Europe regarding the limit of the presence of hydrocarbons are contrasting. Mexico divides into three categories as NOM-138 establishes the maximum permissible number of hydrocarbons in the soil (mg/kg). Agriculture goes up 200–3000, residential and recreational 200–3000, and industrial from 500 to 6000. By Royal Decree 9/2005, Spain establishes the limits for hydrocarbons as a contaminated area if the concentration is 50 mg/kg, but the maximum limit is 5000 mg/kg, although, for this rate, there must be a soil analysis. The limit only represents for the contaminated area in general. The legal limit for hydrocarbons, in the US, is 500 ppm in workplaces, although there is no federal regulation for hydrocarbons.

7. Conclusion

Bioremediation is the solution to reverse contamination problems. Maintaining control of *in situ* conditions remains challenging to achieve 100% passive environmental recovery. Nowadays, more attention is paid to evaluating cost—the benefits of these approaches. Once the analytical methods not only follow the bulk contaminant metabolization but also monitor the possible toxic byproducts in the site and are standardized, consider the insoluble nature of the contaminant. It will be possible to translate all the scientific advances in the field to the sites to recuperate.

Conflict of interest

The authors declare no conflict of interest.

Author details

Juan Cabral-Miramontes[1], Pamela Dorantes-Alvarado[2]
and Elva Teresa Aréchiga-Carvajal[2]*

1 Faculty of Chemical Sciences (Durango Unit), University Juarez State of Durango, Victoria de Durango, Mexico

2 Biological Sciences Faculty, Genetic Manipulation Unit, Laboratory of Mycology and Phytopathology, Autonomous University of Nuevo Leon, San Nicolás of the Garza, Nuevo León, Mexico

*Address all correspondence to: elva.arechigacr@uanl.edu.mx

References

[1] Koolivand A, Abtahi H, Parhamfar M, Saeedi R, Coulon F, Kumar V, et al. The effect of petroleum hydrocarbons concentration on competition between oil-degrading bacteria and indigenous compost microorganisms in petroleum sludge bioremediation. Environmental Technology and Innovation. 2022;**26**:102319

[2] Fox CP, Whiteside JH, Olsen PE, Grice K. Flame out! End-Triassic mass extinction polycyclic aromatic hydrocarbons reflect more than just fire. Earth and Planetary Science Letters. 2022;**584**:117418

[3] Mishra S, Hora S, Mishra R, Kanaujia PK. Bioremediation-based microorganisms to break down pollutants decelerate due to climate change. In: Plant Stress Mitigators: Action and Application. Singapore: Springer Nature Singapore; 2022. pp. 125-143

[4] Kaur G, Kaur G, Krol M, Brar SK. Unraveling the mystery of subsurface microorganisms in bioremediation. Current Research in Biotechnology. 2022;**4**:302-308

[5] Dar A, Naseer A. Recent applications of bioremediation and its impact. Hazardous Waste Management. London, UK: IntechOpen; 2022. p. 49

[6] León HT, San Martín YB, Pérez AP, Silva RR, Díaz SA. Diseño y caracterización de un consorcio bacteriano para la degradación de ripios de perforación base diésel. Revista CENIC Ciencias Químicas. 2022;**53**(2):102-112

[7] Ahmed F, Fakhruddin ANM. A review on environmental contamination of petroleum hydrocarbons and its biodegradation. International Journal of Environmental Sciences & Natural Resources. 2018;**11**(3):1-7. DOI: 10.19080/IJESNR.2018.11.555811

[8] Salazar Carranza LA, Hinojoza Guerrero MM, Acosta Gaibor MP, Escobar Torres AF, Scrich Vázquez AJ. Caracterización, clasificación y usos de las enzimas lipasas en la producción industrial. Revista Cubana de Investigaciones Biomedicas. 2020;**39**(4)

[9] Martini VP, Krieger N, Glogauer A, Souza EM, Iulek J. Structure solution and analyses of the first true lipase obtained from metagenomics indicate potential for increased thermostability. New Biotechnology. 2019;**53**:65-72

[10] Bustos AS, Håkansson A, Linares-PasténJA,PenarrietaJM,NilssonL. Interaction between phenolic compounds and lipase: the influence of solubility and presence of particles in the IC50 value. Journal of Food Science. 2018;**83**(8):2071-2076

[11] SSSA. Soil Science Society of America. 2022. Available from: https://www.soils.org/

[12] Leahy JG, Colwell RR. Microbial degradation of hydrocarbons in the environment. Microbiological Reviews. 1990;**54**(3):305-315. DOI: 10.1128/mr.54.3.305-315.1990

[13] Al-Hawash AB, Dragh MA, Li S, Alhujaily A, Abbood HA, Zhang X, et al. Principles of microbial degradation of petroleum hydrocarbons in the environment. Egyptian Journal of Aquatic Research. 2018;**44**(2):71-76. Available from. https://www.sciencedirect.com/science/article/pii/S1687428518300244

[14] Varjani SJ. Microbial degradation of petroleum hydrocarbons. Bioresource

Technology. 2017;**223**:277-286. Available from : https://www.sciencedirect.com/science/article/pii/S0960852416314432

[15] Wilkes H, Buckel W, Golding BT, Rabus R. Metabolism of hydrocarbons in n-alkane-utilizing anaerobic bacteria. Journal of Molecular Microbiology and Biotechnology. 2016;**26**(1-3):138-151. Available from: https://www.karger.com/DOI/10.1159/000442160

[16] Wang W, Shao Z. Enzymes and genes involved in aerobic alkane degradation. Frontiers in Microbiology. 2013;**4**:116

[17] Benedek T, Szentgyörgyi F, Szabó I, Kriszt B, Révész F, Radó J, et al. Aerobic and oxygen-limited enrichment of BTEX-degrading biofilm bacteria: dominance of Malikia versus Acidovorax species. Environmental Science and Pollution Research International. 2018;**25**(32):32178-32195. DOI: 10.1007/s11356-018-3096-6

[18] Olajire AA, Essien JP. Aerobic degradation of petroleum components by microbial consortia. Journal of Petroleum & Environmental Biotechnology. 2014;**5**(5):1. DOI: 10.4172/2157-7463.1000195

[19] Abbasian F, Lockington R, Mallavarapu M, Naidu R. A comprehensive review of aliphatic hydrocarbon biodegradation by bacteria. Applied Biochemistry and Biotechnology. 2015;**176**(3):670-699. DOI: 10.1007/s12010-015-1603-5

[20] Meckenstock RU, Boll M, Mouttaki H, Koelschbach JS, Cunha Tarouco P, Weyrauch P, et al. Anaerobic degradation of benzene and polycyclic aromatic hydrocarbons. Journal of Molecular Microbiology and Biotechnology. 2016;**26**(1-3):92-118. Available from: https://d-nb.info/1247430596/34

[21] Rojo F. Enzymes for aerobic degradation of alkanes. Handbook of Hydrocarbon and Lipid Microbiology. 2010;**2**:781-797

[22] Imron MF, Kurniawan SB, Ismail NI, Abdullah SRS. Future challenges in diesel biodegradation by bacteria isolates: a review. Journal of Cleaner Production. 2020;**251**:119716

[23] Li X, Li H, Qu C. A review of the mechanism of microbial degradation of petroleum pollution. In: IOP Conference Series: Materials Science and Engineering, The 5th Annual International Conference on Material Engineering and Application (ICMEA 2018), Wuhan, China, 14-16 December 2018; vol. 484. Institute of Physics Publishing: Bristol, UK, 2019. pp. 1-4

[24] Khalid FE, Lim ZS, Sabri S, Gomez-Fuentes C, Zulkharnain A, Ahmad SA. Bioremediation of diesel contaminated marine water by bacteria: a review and bibliometric analysis. Journal of Marine Science and Engineering. 2021;**9**(2):155

[25] Elumalai P, Parthipan P, Karthikeyan OP, Rajasekar A. Enzyme-mediated biodegradation of long-chain n-alkanes (C32 and C40) by thermophilic bacteria. 3 Biotech 2017;**7**:1-10

[26] Gregson BH, Metodieva G, Metodiev MV, Golyshin PN, McKew BA. Differential protein expression during growth on medium versus long-chain alkanes in the obligate marine hydrocarbon-degrading bacterium Thalassolituus oleivorans MIL-1. Frontiers in Microbiology. 2018;**9**:3130

[27] Medić A, Lješević M, Inui H, Beškoski V, Kojić I, Stojanović K, et al. Efficient biodegradation of petroleum n-alkanes and polycyclic aromatic hydrocarbons by polyextremophilic

Pseudomonas aeruginosa san ai with multidegradative capacity. RSC Advances. 2020;**10**:14060-14070

[28] Moreno R, Rojo F. Enzymes for aerobic degradation of alkanes in bacteria. In: Aerobic Utilization of Hydrocarbons, Oils, and Lipids. Basel, Switzerland: Springer International Publishing; 2017. pp. 1-25

[29] Sarmah N, Revathi D, Sheelu G, Yamuna Rani K, Sridhar S, Mehtab V, et al. Recent advances on sources and industrial applications of lipases. Biotechnology Progress. 2018;**34**(1):5-28

[30] Laachari F, El Bergadi F, Sayari A, Elabed S, Mohammed I, Harchali EH, et al. Biochemical characterization of a new thermostable lipase from Bacillus pumilus strain/[Bacillus pumilus suşundan elde edilen yeni termostabil lipazın biyokimyasal karakterizasyonu]. Turkish Journal of Biochemistry. 2015;**40**(1):8-14

[31] Bose A, Keharia H. Production, characterization and applications of organic solvent tolerant lipase by Pseudomonas aeruginosa AAU2. Biocatalysis and Agricultural Biotechnology. 2013;**2**(3):255-66

[32] Jo JC, Kim SJ, Kim HK. Transesterification of plant oils using Staphylococcus haemolyticus L62 lipase displayed on Escherichia coli cell surface using the OmpA signal peptide and EstAβ8 anchoring motif. Enzyme and Microbial Technology. 2014;**67**:32-39

[33] Işık C, Saraç N, Teke M, Uğur A. A new bioremediation method for removal of wastewater containing oils with high oleic acid composition: Acinetobacter haemolyticus lipase immobilized on eggshell membrane with improved stabilities. New Journal of Chemistry. 2021;**45**(4):1984-1992. Available from: https://pubs.rsc.org/en/content/articlelanding/2021/nj/d0nj05175f/unauth

[34] Panizza P, Syfantou N, Pastor FI, Rodriguez S, Diaz P, Acidic lipase L ip I. 3 from a P seudomonas fluorescens-like strain displays unusual properties and shows activity on secondary alcohols. Journal of Applied Microbiology. 2013;**114**(3):722-732

[35] Zheng C. Growth characteristics and enzyme production optimization of lipase producing strain. IOP Conference Series: Earth and Environmental Science. 2018;**108**:042087. DOI: 10.1088/1755-1315/108/4/042087

[36] Wancura JHC, Rosset DV, Mazutti MA, Ugalde GA, de Oliveira JV, Tres MV, et al. Improving the soluble lipase-catalyzed biodiesel production through a two-step hydroesterification reaction system. Applied Microbiology and Biotechnology. 2019;**103**(18):7805-7817. DOI: 10.1007/s00253-019-10075-y

[37] Chandra P, Enespa SR, Arora PK. Microbial lipases and their industrial applications: a comprehensive review. Microbial Cell Factories. 2020;**19**(1):169. DOI: 10.1186/s12934-020-01428-8

[38] Pang SM, Le S, Kwiatkowski AV, Yan J. Mechanical stability of αT-catenin and its activation by force for vinculin binding. Molecular Biology of the Cell. 2019;**30**(16):1930-1937

[39] Nguyen HH, Kim M. An overview of techniques in enzyme immobilization. Applied Science and Convergence Technology. 2017;**26**(6):157-163

[40] Albayati SH, Masomian M, Ishak SNH, Mohamad Ali MSB, Thean AL, Mohd Shariff FB, et al. Main structural targets for engineering

lipase substrate specificity. Catalysts. 2020;**10**(7):747. Available from: https://www.mdpi.com/2073-4344/10/7/747

[41] Zhao J, Ma M, Zeng Z, Yu P, Gong D, Deng S. Production, purification and biochemical characterisation of a novel lipase from a newly identified lipolytic bacterium Staphylococcus caprae NCU S6. Journal of Enzyme Inhibition and Medicinal Chemistry. 2021;**36**(1):248-256. Available from:. DOI: 10.1080/14756366.2020.1861607

[42] Mohamad NR, Marzuki NH, Buang NA, Huyop F, Wahab RA. An overview of technologies for immobilization of enzymes and surface analysis techniques for immobilized enzymes. Biotechnology and Biotechnological Equipment. 2015;**29**(2):205-220

[43] Choi JM, Han SS, Kim HS. Industrial applications of enzyme biocatalysis: current status and future aspects. Biotechnology Advances. 2015;**33**(7):1443-1454

[44] Vršanská M, Voběrková S, Jimenez Jimenez AM, Strmiska V, Adam V. Preparation and optimisation of cross-linked enzyme aggregates using native isolate white rot fungi Trametes versicolor and Fomes fomentarius for the decolourisation of synthetic dyes. International Journal of Environmental Research and Public Health. 2018;**15**(1):23

[45] Reis CL, Sousa EY, Serpa JD, Oliveira RC, Santos JC. Design of immobilized enzyme biocatalysts: drawbacks and opportunities. Química Nova. 2019;**42**:768-783

[46] Ye J, Chu T, Chu J, Gao B, He B. A versatile approach for enzyme immobilization using chemically modified 3D-printed scaffolds. ACS Sustainable Chemistry & Engineering. 2019;**7**(21):18048-18054

[47] Jorge G-B, Ricardo M-MV, del Monte Martínez Alberto. Las lipasas: enzimas con potencial para el desarrollo de biocatalizadores inmovilizados por adsorción interfacial. Revista Colombiana de Biotecnología. 2010;**12**(1):113-140. Available from: http://www.scielo.org.co/scielo.php?script=sci_arttext&pid=S0123-34752010000100013&lng=en

[48] Brown LR. Microbial enhanced oil recovery (MEOR). Current Opinion in Microbiology. 2010;**13**:316-320

[49] Van Beilen JB, Funhoff EG. Alkane hydroxylases involved in microbial alkane degradation. Applied Microbiology and Biotechnology. 2007;**74**:13-21

[50] Bihari Z et al. Functional analysis of long-chain n-alkane degradation by Dietzia spp. FEMS Microbiology Letters. 2011;**316**:100-107

[51] Wang XB et al. Degradation of petroleum hydrocarbons (C6-C40) and crude oil by a novel Dietzia strain. Bioresource Technology. 2011;**102**:7755-7761

[52] Nie Y, Chi CQ, Fang H, Liang JL, Lu SL, Lai GL, et al. Diverse alkane hydroxylase genes in microorganisms and environments. Scientific Reports. 2014;**4**(1):1-11

[53] Ji Y, Mao G, Wang Y, Bartlam M. Structural insights into diversity and n-alkane biodegradation mechanisms of alkane hydroxylases. Frontiers in Microbiology. 2013;**4**:58

[54] Grady SL, Malfatti SA, Gunasekera TS, Dalley BK, Lyman MG, Striebich RC, et al. A comprehensive

multi-omics approach uncovers adaptations for growth and survival of Pseudomonas aeruginosa on n-alkanes. BMC Genomics. 2017;**18**(1):1-19

[55] Douglas GS, McCarthy KJ, Dahlen DT, Seavey JA, Steinhauer WG, Prince RC, et al. The use of hydrocarbon analyses for environmental assessment and remediation. Journal of Soil Contamination. 1992;**1**:197e216

[56] Mansur AA, Taha M, Shahsavari E, Haleyur N, Adetutu EM, Ball AS. An effective soil slurry bioremediation protocol for the treatment of Libyan soil contaminated with crude oil tank bottom sludge. International Biodeterioration and Biodegradation. 2016;**115**:179e185

[57] Wu M, Li W, Dick WA, Ye X, Chen K, Kost D, et al. Bioremediation of hydrocarbon degradation in a petroleum-contaminated soil and microbial population and activity determination. Chemosphere. 2017;**169**:124e130

[58] Booth AM, Sutton PA, Lewis CA, Lewis AC, Scarlett A, Chau W, et al. Unresolved complex mixtures of aromatic hydrocarbons: thousands of overlooked persistent, bioaccumulative, and toxic contaminants in mussels. Environmental Science & Technology. 2007;**41**:457e464

[59] Tran TC, Logan GA, Grosjean E, Ryan D, Marriott PJ. Use of comprehensive two-dimensional gas chromatography/time-of-flight mass spectrometry for the characterization of biodegradation and unresolved complex mixtures in petroleum. Geochimica et Cosmochimica Acta. 2010;**74**:6468e6484

[60] Pena M, Casais M, Mejuto M, Cela R. Optimization of the matrix solid-phase dispersion sample preparation procedure for analysis of polycyclic aromatic hydrocarbons in soils: comparison with microwave-assisted extraction. Journal of Chromatography A. 2007;**1165**:32e38

[61] Akhlaq MS. Polycyclic aromatic hydrocarbons in crude oil-contaminated soil: a two-step method for the isolation and characterization of PAHs. Environmental Science and Pollution Research. 1997;**4**:217e222

[62] Yu B, Tian J, Feng L. Remediation of PAH polluted soils using a soil microbial fuel cell: influence of electrode interval and role of microbial community. Journal of Hazardous Materials. 2017;**336**:110e118

[63] Wu ML, Nie MQ, Wang XC, Su JM, Cao W. Analysis of phenanthrene biodegradation by using FTIR, UV and GC-MS. Spectrochimica Acta, Part A: Molecular and Biomolecular Spectroscopy. 2010;**75**:1047e1050

[64] Qi Y-B, Wang C-Y, Lv C-Y, Lun Z-M, Zheng C-G. Removal capacities of polycyclic aromatic hydrocarbons (PAHs) by a newly isolated strain from oilfield produced water. International Journal of Environmental Research and Public Health. 2017;**14**:215

[65] Selifonov SA, Chapman PJ, Akkerman SB, Gurst JE, Bortiatynski JM, Nanny MA, et al. Use of 13C nuclear magnetic resonance to assess fossil fuel biodegradation: fate of [1-13C] acenaphthene in creosote polycyclic aromatic compound mixtures degraded by bacteria. Applied and Environmental Microbiology. 1998;**64**:1447e1453

[66] Rodgers-Vieira EA, Zhang Z, Adrion AC, Gold A, Aitken MD. Identification of anthraquinone-degrading bacteria in soil contaminated with polycyclic aromatic hydrocarbons. Applied and Environmental Microbiology. 2015;**81**:3775e3781

[67] Imam A, Suman SK, Ghosh D, Kanaujia PK. Analytical approaches used in monitoring the bioremediation of hydrocarbons in petroleum-contaminated soil and sludge. TrAC Trends in Analytical Chemistry. 2019;**118**:50-64

[68] Singh RS, Singh T, Pandey A. Microbial enzymes—an overview. In: Singh RS, Singhania RR, Pandey A, Larroche C, editors. Advances in Enzyme Technology. Elsevier; 2019. pp. 1-40

[69] NORMA Oficial Mexicana NOM-138-SEMARNAT/SSA1-2012, Límites máximos permisibles de hidrocarburos en suelos y lineamientos para el muestreo en la caracterización y especificaciones para la remediación. https://www.dof.gob.mx/nota_detalle.php?codigo=5313544&fecha=10/09/2013#gsc.tab=0

[70] NORMA Oficial Mexicana NOM-035-STPS-2018, Factores de riesgo psicosocial en el trabajo-Identificación, análisis y prevención. https://www.dof.gob.mx/nota_detalle.php?codigo=5541828&fecha=23/10/2018#gsc.tab=0

[71] NORMA Oficial Mexicana, NOM-147-SEMARNAT/SSA1-2004, Que establece criterios para determiner las concentraciones de remediación de suelos contaminados por arsénico, bario,berilio, cadmio, cromo hexavalente, mercurio, níquel, plata, plomo, selenio, talio y/vanadio

[72] NORMA Oficial Mexicana NOM-052-SEMARNAT-2005, Que establece las características, el procedimiento de identificación, clasificación y los listados de los residuos peligrosos. https://www.dof.gob.mx/normasOficiales/1055/SEMARNA/SEMARNA.htm

[73] Programa Nacional de Remediacion de Sitios Contaminados 2021-2024, publicado 05/11/2021 https://www.dof.gob.mx/nota_detalle.php?codigo=5634656&fecha=05/11/2021#gsc.tab=0

[74] Us Epa O. Green remediation best management practices: bioremediation. 2015; Available from: https://www.epa.gov/remedytech/green-remediation-best-management-practices-bioremediation

[75] Epa.gov. Available from: https://www.epa.gov/sites/default/files/2015-06/documents/epa-524.2.pdf

[76] EPA-542-F-16-005. Best Practices for Environmental Site Management: Recommended Contents of a Groundwater Monitoring Report. 2016. Available from: https://semspub.epa.gov/work/HQ/500024623.pdf

[77] Ley 22/2011, de 28 de julio, de residuos y suelos contaminados. https://www.boe.es/buscar/act.php?id=BOE-A-2011-13046&p=20220409&tn=0

[78] Real Decreto 9/2005, de 14 de enero, por el que se establece la relación de actividades potencialmente contaminantes del suelo y los criterios y estándares para la declaración de suelos contaminados. https://www.boe.es/buscar/doc.php?id=BOE-A-2005-895

[79] Real Decreto 664/1997, de 12 de mayo, BOE n° 124, de 24 de mayo https://www.boe.es/buscar/act.php?id=BOE-A-2011-13046&p=20220409&tn=0

Chapter 5

Biotransformation of Metal-Rich Effluents and Potential Recycle Applications

Suzan P. Vasconcellos, André Paganotti, Vitor G. Vital, Lidiane M. Santos Lima, Giovanna S.M. Paiva, L. Furlaneto de Lima, Enrique Moreira, Leticia O. Sousa, Guilherme G. Guerini, Vinicius T. Santos, Flavia G. Lobo, Márcio R. Silva, Diogo S. Pellosi and Ricardo A.G. Silva

Abstract

In this chapter, it was introduced about the metallurgic effluents, and their potential to be converted into some feasible coproducts for industries. Some possibilities to introduce circular economy in the context of metallurgic effluents, and in the same way, some techniques to promote bioremediation using microorganisms and products from them were also described. Reported studies, as well as some perspectives to use metal-rich effluents in agriculture and soil quality improvement, were also shown. Copper effluents were kept as the main candidate for sustainable use, as a potentially interesting material for circular economy approaches.

Keywords: copper, metals, nanotechnology, microorganisms, agriculture

1. Introduction

Since the Stone Age, minerals have played an important role in the development of civilization, with mining activity being present from the beginning to the present days of society [1]. In fact, humanity is currently consuming mineral resources at an unprecedented rate of 70 Gt/year, resulting in the highest per-capita levels of resource consumption in history [2]. The excessive use of mineral resources results in a substantial generation of different kinds of wastes including liquid effluents.

These are wastes discharged from industrial, commercial, or domestic facilities and usually may contain many types of pollutants, including chemicals, heavy metals, salts, and pathogens, which can have harmful effects on human, animal, and environmental health. Regarding its end destination, effluents can be sent to treatment plants located right near the source of the effluent, usually private facilities, or common treatment plants farther from the generation point.

Although many effluents can be fully treated, those containing metallics or persistent organic pollutants pose many challenges to the current remediation techniques. Due to the challenges faced in treating such waste and its high cytotoxicity, good practices in industry consist of an "*in situ*" treatment, avoiding its relocation and transportation [3].

As regards the treatment of metal-rich effluents, traditional physicochemical methods, such as chemical precipitation and adsorption reactions, have been applied for many years to remove heavy metals from wastewater. However, these methods could have some problematic limitations, including high costs, generation of great amounts of sludge, as well as low efficiency in the removal of contaminants [4, 5]. Therefore, it is important to separate and preferably, as mentioned, treat these metal effluents *in situ*, since traditional techniques are not suited for treating heavy metal-based liquid wastes. For example, the sludge that is generated in the common effluent treatment is poisoned by the metals and a prolonged exposure to this effluent may induce resistance in the microbiome of the treatment plant by metal absorption in the long run. The observation of these processes leads to a new field of remediation, the so-called bioremediation.

Bioremediation has emerged as a promising alternative for the treatment of metal-rich effluents due to its effectiveness, low cost, and eco-friendly nature [6]. Bioremediation could be defined as the use of living organisms, such as bacteria, fungi, algae, and plants, to remove or transform hazardous substances from the environment [6, 7]. Microorganisms have evolved various measures to face heavy-metal stress, via processes such as transport across the cell membrane, biosorption

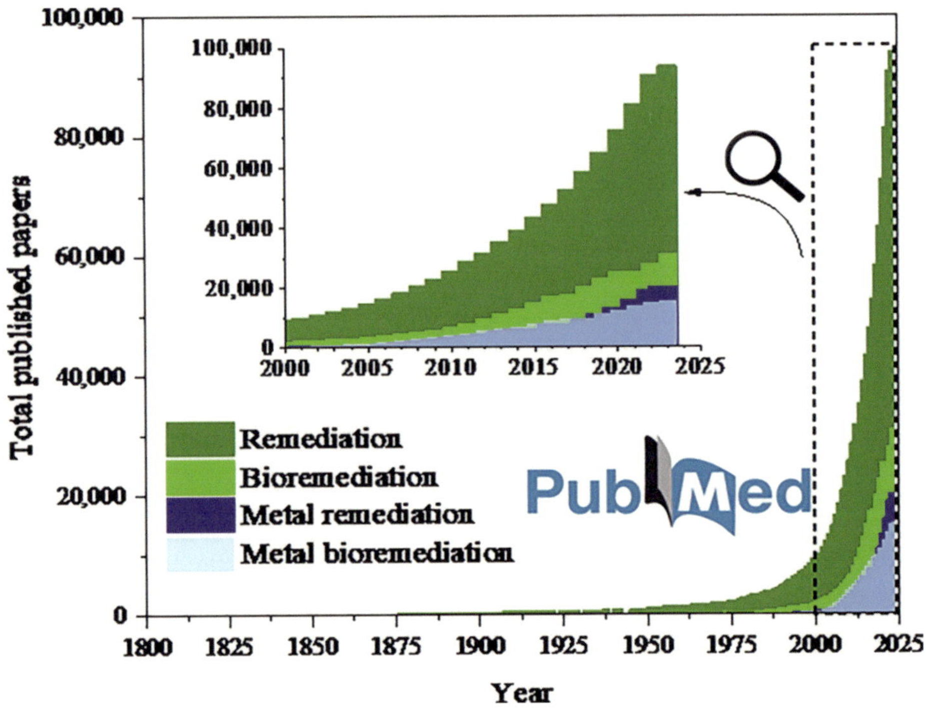

Figure 1.
Published papers registered in PubMed by year in the areas of remediation, bioremediation, metal remediation, and metal bioremediation. Data shown until April of 2023.

to cell walls, and entrapment in extracellular capsules, precipitation, complexation, and oxidation-reduction reactions [8]. Since many microorganisms already present mechanisms to intake and use metals in their metabolism as micronutrients, these organisms can be used to accumulate large amounts of heavy metals from the mining and metallurgical industries [9]. In nature, this intricate relationship is responsible for the metal biogeochemistry leading to the mobilization or immobilization of metal species [10]. First cited in 1969 [11], metal bioremediation has gained significant attention in the past 20 years, being responsible for approximately half of the total bioremediation papers published, as can be seen in **Figure 1**.

This chapter focuses on the score during 2003–2023 when the bioremediation evolved from theory to application. To showcase the fast-paced development of this technology, from 1969 to 2002, almost 800 papers were published, and in the past two decades however, more than 14,000 papers were published citing metal bioremediation. This shows the expansion of metal bioremediation and its relevance for metal industries' circular economy aiming at eco-friendly approaches in metal-rich effluents' treatment.

2. Early years (2003–2010)

As mentioned earlier, at the beginning of the twenty-first century, bioremediation processes were already known but faced practical limitations. Before this period, the relationships between many metal species and microorganisms have already been established, and microorganisms have been suggested for remediating metals from liquid wastes.

The bioremediation process can occur through different pathways, such as biosorption, bioaccumulation, biomineralization, and biodegradation. Biosorption involves the passive binding of heavy metals onto the surface of microbial cells that do not even need to be alive for this process to occur. Bioaccumulation involves the active uptake of heavy metals into the cells. Biomineralization involves the precipitation of heavy metals as insoluble compounds, while biodegradation is related to the breakdown of heavy-metal compounds into less toxic forms. The chosen approach may change considering the effluent metal concentration and type [4].

Some of the most common metals found in effluents include arsenic, lead, cadmium, chromium, copper, nickel, silver, and zinc [12, 13]. These elements are present in a plethora of effluents originated from different types of industries. Some industries produce effluents with a prevalent metal, usually coming from its main substrate. For example, Pb is commonly used in batteries and ammunition; Cd is also commonly used in batteries and coatings, while Cu is normally used in electrical wiring and plumbing, as well as other applications. However, some industries can produce some effluents with a mix of metals as, for example, mining industries, electroplating facilities, and paint manufactures. This mixture of different metals usually shows a difficult situation for the bioremediation processes of these industrial effluents.

In fact, metal effluents can be produced at many concentration ranges, but the traditional methods fail to be economically viable at low concentrations that usually range from 1 to 100 mg/L [14], thus increasing costs of separation and processing. On the other hand, at low concentrations, bioremediation processes stand out since the microorganisms can easily adapt to the prevalent conditions if compared to traditional physicochemical methods.

The first form of bioremediation of metals can be passive biosorption using biomass. These first efforts were also interested in the recovery of the metal species since environmental regulations were starting to compel industries to shift to cleaner methods [15]. At this time, the high metal-binding capability of many biological materials had already been described in the previous decade (in the 1990s) including algae, bacteria, and fungi, many of them, isolated from food industries. Many of these biosorption processes did not require the use of whole cells of microorganisms, using the presence of metal-binding molecules into the biomass to capture the metals from the effluent. The described process could not be influenced by the metabolic cycle of the biomass and it is commonly referred to as "passive uptake." This uptake could be analyzed using biosorption isotherm curves, evaluating the influence of the sorption process conditions [8]. These studies were focused on synthetic effluents, treating them with prepared biomass material, often using chemicals to increase the absorption into the biomass. These reported biosorption studies of a single metal on dead biomass, for example, Ni sorption onto dead mass presented sorption of 0.3% in low concentration synthetic effluents [8].

Despite being a kickstart to the technology, this process did not deal well under mixed metal effluents, usually disrupting the process, which hampered industrial applications [8, 13]. In addition, the biomass needed to be treated with chemicals, such as sodium hydroxide, detergents, formaldehyde, dimethyl sulfoxide, etc. One solution for this problem was the selection of microbial strains isolated from contaminated sites. The idea was that these strains, probably, already presented the capability to metabolize metal and promote bioremediation at some level, showing some adaptability to a wide range of concentrations [16, 17]. When the pure biosorptive metal removal was not viable, a consortium of microorganisms could be used, combining biosorption with other techniques, including bioaccumulation and mineralization [8].

It was also in these early years (2003–2010) that researchers started to establish that living and growing cells could uptake (the so-called bioaccumulation) great amounts of metal ions in solution in an extension even greater than that of the described biomass [8, 18]. These studies showed the capability of the metal to be internalized via an active uptake during the organism growth, increasing the metal uptake by joining the passive biosorption that initially happened. These changes resulted in metal concentrations to be higher inside the living cells than the biosorbed ones, increasing the efficiency of the bioremediation process.

Only at the end of the decade, high removal efficiencies were observed during the bioremediation of mixed metal effluents, such as mining effluents with bioremediation processes reaching 91, 96, and 99% of removal for Fe, Cu, and Al, respectively [19]. And in electroplating solutions using yeast (*Saccharomyces cerevisiae*) reaching 92, 92, and 87% of removal for Cu, Cr(VI), and Ni [20]. Both studies used subproducts of the food industry applying a cheap biomass originated from the microbial growth in the production of cheese [19] and wine [20]. From these studies, fungal biomass gained an increasing interest for removing metals. Fungal biomass is a subproduct of many industrial processes and fermentations, presenting a wide variety of morphologies [9]. The abundance of fungal mass available and the fact that usually fungi are already adapted to lower pH's made them ideal candidates for bioremediation of metal-rich effluents. Other studies showed some strains of *Aspergillus niger* have higher Ni biosorption capacities, if compared to dead pretreated biomass [8]. Also, the same fungal strains of *A. niger* showed high resistance to lead (II), however chromium (VI) ions caused inhibition of the microorganism that thrived in the presence of copper and lead [9].

To overcome this drawback, researchers obtained new microbial consortia from distinct contaminated matrices. For example, Jiménes-Rodriques et al. [19] collected samples of biomass from the Río Tinto, a river in Spain that presents an unusual acidic pH and elevated concentrations of metals. The obtained microbial consortia from this river were then applied to treat mine drainage by simply regulating the pH in an anaerobic condition, leading to efficiencies in metal removal between 91 and 99%.

It is important to notice the influence of the pH in the process since the conducted experiments by MacHado et al. [20] showed that by adjusting the pH of the treatment, it was possible to, selectively, remove some ions. At very low pH (around 2.0), the yeast surfaces were surrounded by hydroniums, enhancing the interaction of the biomass with Cr (VI). This is a very interesting approach, since using different pH ranges, it was possible to remove chromium ions selectively, avoiding poisoning of the sludge. The efficiency of Cr(IV) removal achieved using heat-inactivated cells of a flocculent strain of *Saccharomyces cerevisiae* was between 90 and 99% [20].

Although present at this time, phytoremediation was considered in its early stages. This technology is based on the growth of plant-based organisms on contaminated sites in a way that the plant can accumulate high quantities of metals. However, the efficiency of waste remediation is intrinsically related to the plant growth rate and the total biomass, making the process slow. In this scenario, the use of fast-growing plants was necessary [15], being the biomass of algae considered as an innovative solution [14].

In his influential review, Kosolapov et al. [21] described the most promising processes involved in the bioremediation of metals in constructed wetlands, i.e., engineered systems designed to mimic the natural processes of a wetland environment for the treatment of wastewater. This review also brought about the possible interactions of the microorganisms and plants present at these wetlands (**Figure 2**).

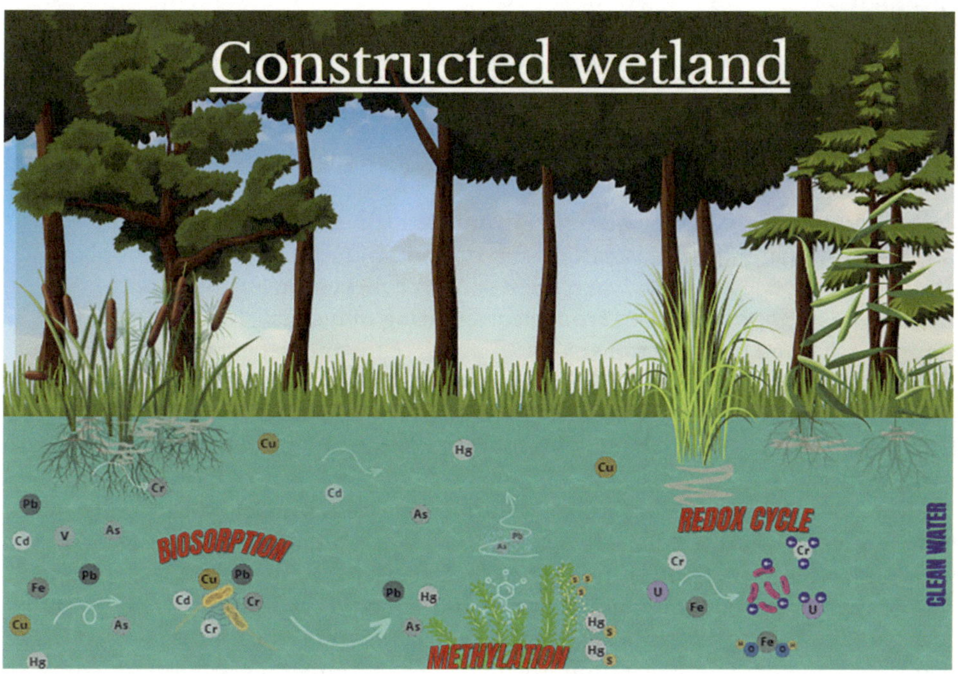

Figure 2.
Illustration based on the review from Kosolapov et al. [21].

These early years of metal bioremediation were summarized by Stasinakis and Thomaidis [22]. This study visited the reports of the biotransformation potential of many metals and metalloids, in experiments involving both, pure and mixed microbial cultures. Marked by great advances in the use of microorganisms in metal remediation, these early years cemented the fundamentals and experimental parameters needed for the implementation of an efficient bioremediation process to real effluents. However, despite the great advance in removal efficiency and elemental speciation, data on the biotransformation in Wastewater Treatment Processes were not enough. Fortunately, it did not take long before these data were enough for the bioremediation process to reach the next level of excellence.

3. Last decade (2010–2020)

This decade was marked by the application of bioremediation processes with high removal rates in real effluents and advances in isolating new microorganisms as alternatives to treat multimetal effluents [23]. For example, in 2013, a strain of *Acinetobacter* sp. was isolated from an effluent treatment plant in New Delhi, India, which showed high tolerance to the Cr (VI), exhibiting a high removal rate of Cr and Ni from a real effluent in an electroplating facility [23]. These chromium-resistant microorganisms used various detoxification strategies to counteract the Cr (VI), including enzymes'/metabolites' biotransformation from Cr (VI) to Cr (III), which is less toxic [24]. It is interesting to note that these resistant strains arise from the failure in treating metal-rich effluents, which indicated high content of these toxic metals in the environment. In another relevant work, Zhao et al. [25] isolated strains of *Sporosarcina saromensis* from offshore sediments in China, capable of tolerating super high Cr (VI) concentrations (~500 mg/l). This strain could remove 50–200 mg of Cr (VI)/L in just 24 h. This finding just demonstrated that with the rise in offshore pollution, tolerant strains have been commonly isolated from offshore and intertidal zones. From this study, more than 50 strains with the ability to tolerate Cr (VI) were isolated [25]. These findings confirmed that the current waste treatment was not adequate and that the environmental contamination by metals became a serious threat [26].

Cr-resistant *Bacillus* sp. was isolated from the effluent from a local tannery in Kanpur, UP, India [27]. Also isolated from the common effluent treatment plant of tannery industries, *Cellulosimicrobium* sp. was collected in Kanpur, India [28]. *Trichoderma viride* was isolated from electroplating industrial sludge [29]. Other examples are listed by Tarekegn et al. [26] and Kapahi and Sachdeva [30]. As well summarized by Tarekegn et al. [26] at the end of the decade: "Autochthonous (indigenous) microorganisms present in polluted environments hold the key to solving most of the challenges associated with biodegradation and bioremediation of polluting substances."

Although not many studies were carried out, it is also important to highlight the use of microbial mixtures. In 2015, Kang et al. [31] isolated four bacterial strains from an abandoned mine in South Korea. These bacteria lived in the soil of the mine and presented excellent Pb bioremediation capabilities. In 2016, Kang et al. [32] mixed the isolated microorganisms in different proportions and their results showed that when compared with a single strain, the mixtures presented higher growth rate, urease activity, and resistance to metals. These factors allowed the mixtures to exhibit considerably higher bioremediation capabilities when compared to single strains.

It was also in this decade that researchers started to better establish the role of microorganisms in the metabolisms of plants. For example, it was found that many endophytic organisms permitted the absorption of metal species [33]. This idea associated with the fact that many biomes were already being exposed to metals raises attention to phytoremediation. The term phytoremediation describes the utilization of plants and their accompanying microorganisms to remediate specific contaminants from soil, sludge, sediments, wastewater, and groundwater, either partially or entirely [34]. The ability of plants and associated microorganisms to accumulate essential metals in their metabolism enables them to accumulate other nonessential metals. It is suitable when the pollutants cover a wide area and when they are within the root zone of the plant [35]. The discovery of hyperaccumulators (plant species that naturally present a very high metal absorption) rekindled the interest in the area [34]. These plants can also be used with plant-growth-promoting bacteria, and these bacteria can improve the metal uptake by turning the metal species in the soil/effluent bioavailable to the plant [36].

Plant-like organisms such as algae also became a very interesting approach. Used mainly as biomass at the beginning of bioremediation, in this decade, the use of living algae marked an emerging trend of phytoremediation. Microalgae possess remarkable biological traits, including high photosynthetic efficiency and a simple structure, enabling them to thrive in challenging environmental conditions, such as the presence of metals [37]. In comparison with the traditional use of plants in phytoremediation, the use of algae presented lower toxicity constraint, high growth rate, and the formation of value-added products such as biofuels and fertilizer [37], bioremediating the metals and promoting carbon capture and circular economics. This new technique would gain even more attention in the next years, due to varied tolerance and specific responses as well as high metal bioaccumulation. Further advances in genetic engineering, metal immobilization techniques, algae pretreatment strategies, and integration with other emerging technologies are needed to fully unlock the potential of phytoremediation [37].

4. Recent years (2020–2023)

In recent years, the isolation of microorganisms continues to take place [38–40] and the intervening new technologies in genetic engineering [41] and omics sciences [42] are making new processes to remediate effluents the most viable solutions. Phytoremediation continued to grow and show many interesting results for metal bioremediation and even dye remediation [4]. Therefore, with the exponential growth of bioremediation studies (see **Figure 1**) using the most diverse kinds of microorganisms and approach based on real-life (multimetal) effluents, biotransformation of metal-rich wastewater is now initiating a new moment in which large-scale processes are close to reaching industrial application. Besides that, even the material produced by the bioremediation can be further utilized, as demonstrated by Amim et al. that showed the importance of the products generated by the biomass from wastewater treatment [43]. In addition, another disadvantage of bioremediation is attributed to its relative slowness when compared to physicochemical methods [44]. However, it is a green approach that presents higher capability and a lower cost that makes bioremediation the best choice aiming at sustainable development following circular economics rules.

Additionally, new approaches based on the union between traditional physicochemical methods and biological processes have been developed. Bio-electro-Fenton is a good example of such type of innovation. In this process, both organic molecules and metal ions are remediated by the microorganisms, while the traditional electro-Fenton process breaks down the persistent organic pollutants (POPs) into intermediates that can be metabolized. At the same time, when the degradation of pollutants to innocuous substances is carried out both by the traditional and biological processes, there is a synergy that reduces the costs of process operation. This is recent in the literature but results already indicate its great capacity for practical application, increasing the efficiency of the process and reducing the electro-intensive costs of the traditional electro-Fenton process [45, 46]. This is an important advance since persistent organic pollutants are an emerging problem in recent decades due to the increased consumption of medications and antibiotics and the fact that they are not 100 percent metabolized, increased fractions of these pollutants end up in the conventional sewage network [47, 48]. These substances present low biodegradability and high risk for human toxicity being generated during industrial production processes, especially those of chemical and pharmaceutical input production.

To degrade POPs, aggressive physicochemical methods are required due to the recalcitrant nature of these pollutants. The so-called advanced oxidative processes can oxidize most classes of recalcitrant pollutants. The Fenton process is an example of this class of treatment and consists in the production of hydroxyl radicals, highly oxidizing species (Eo = 2.80 V), capable of degrading practically any organic substance [49, 50]. This process is very old, being described in 1894 [51], and presents the aforementioned drawbacks, high cost, high sludge generation, etc.

By means of electrocatalysis [52] and photocatalysis [53], both electro-Fenton and photo-Fenton processes were proposed to decrease the Iron II addition by regenerating it through electricity or photon incidence. These processes, once the cutting edge of physicochemical processes, are being now united with biological remediation process to simultaneously treat persistent pollutants and present low cost and impact.

Bio-photo-Fenton [54] and bio-electro-Fenton [46] emerged as new frontiers and Colombo et al. [55] showed the reduction of pollutants in landfill leachate by combining photo-Fenton and biological processes. Silva et al. [56] observed an improvement in the biodegradability of leachate from an aerated lagoon using photo-Fenton powered by solar energy combined with a biologically activated sludge process. And Sirtori et al. [57] treated a real pharmaceutical wastewater by a solar photo-Fenton/biotreatment.

5. After treatment: reusing bioremediated metals in circular economy

Despite the low cost and high efficiency of remediation metals from effluents and soils, the biomass left still concentrates these metal species. Yes, the metal species are less toxic due to the bioremediation process, but at the current metal effluent generation rate, new solutions to further processes involving these metals are needed. The bioremediation process, as discussed, works very well in concentrations too low for traditional chemicals, allowing a low-cost and high-volume metal removal. After the remediation process, however, the metal is highly concentrated in the biomass making the chemical process to transform the metal viable. Therefore, in some cases, bioremediation serves as a pre-concentration step.

However, it is important to note that depending on the bioremediation process used, the metal may be accumulated in different ways inside the biomass. In the biosorption method, the metal can be simply desorbed from the biomass by changing the pH or the temperature of the biomass, or just by washing it with pure water [3]. In the case of bioremediation using growing cells, the metal can be fixed by different metabolic pathways of the bioremediation organism. As an example, the metal can be sequestrated in cell organelles, chelated, or even form ionic bond with the cell membrane [37]. Although some metals present preferential uptakes, this pathway can change from microorganisms to microorganisms, even more so in resistant microorganisms that develop specialty routes to intake metal ions [40]. An example can be *Pseudomonas* sp., depending on its strain, it can bioremediate Ni by accumulating in cytoplasm or in their extracellular polymeric substance [8]. This can be further complicated in consortia when different species concentrate metals by different pathways for the same metal [8, 30].

When simple leaching is possible, the acidic leaching solution removes the biosorbed metal from the adsorbent, leading to a concentrated metal solution that can be recovered by traditional methods such as chelation, reduction, or electrolysis. Moreover, it is also possible to integrate an acidophilic bioleaching with subsequent precipitation of the metals as insoluble sulfides by sulfate-reducing bacteria [8]. In the case of biomasses that are more difficult to leach the metals, usually the lysis of the cells is required. In this case, this can be promoted purely by acid leaching or heat treatment, upcycling the final biomass. A recent advance is the possibility of using simple heating or hydrothermal treatment to produce metal nanoparticles from metal released from the biomass, as demonstrated by Goswami et al. [58]. Other recent advances are related to the use of such metallic nanoparticles in agribusiness, as explored below.

5.1 Metallic nanoparticle development from biomass

With the increase in bottom-up nanoparticle synthesis, metal-rich biomasses are ideal candidates for the green generation of metallic nanoparticles. This type of nanoparticle production is based on the reduction of metal ions to produce metallic or metal oxide nanoparticles. In this context, Cu, Zn, and Ni, and noble metals, such as Ag and Au, present a very interesting opportunity, as bioremediation of these metals is already very well established with high recovery rates, and their nanoparticles are also a very well-established high-technology product [59–61].

This approach, called biogenic synthesis, is an economical and environmentally friendly alternative to chemical and physical approaches for the nanoparticle production [62]. Therefore, biogenic nanoparticles are a growing research field showing promising abilities of microorganisms to produce molecules, by the reduction of metal ions [63]. This can lead to a synergetic relation between the bioremediation process and the production of high-value nanomaterials [58]. In addition, metallic nanoparticles can also be capable of remediating effluents due to their unique properties of high surface-to-volume ratio, large surface area, and enhanced reactivity, which make them highly efficient in capturing and removing metal ions from wastewater [63–65]. Although both biogenic nanoparticles and bioremediation are growing fields, an end-to-end study of removing and recycling metals from effluent are not yet described in consulted literature.

Various types of nanoparticles have been explored for metal effluent remediation, including metallic particles such as iron, copper, nickel, zinc, and their oxides, as well as silica and carbon-based nanoparticles [66]. These nanoparticles can be engineered to have specific surface properties and functionalized with various coatings or ligands to enhance their metal adsorption capacity and selectivity [64]. The mechanism of metal removal using nanoparticles involves several processes, including adsorption, precipitation, ion exchange, and redox reactions. The nanoparticles can adsorb metal ions onto their surfaces, forming complexes through chemical interactions. They can also promote precipitation of other metal ions as metal hydroxides or other insoluble forms.

One advantage of nanoparticle-based remediation is that it can be applied to a wide range of metals, including heavy metals like lead, cadmium, mercury, chromium, and arsenic. Additionally, nanoparticles can be easily synthesized and functionalized, allowing for customization of their properties based on the specific metallic contaminants and wastewater conditions.

5.2 Agribusiness applications

As mentioned, the use of bioremediated metals in the production of nanoparticles can produce high-value products from the remediation processes. One interesting application of these particles is in the agribusiness. Agrochemicals play a crucial role in modern agriculture by providing tools for efficient and sustainable food production. Agrochemicals encompass a wide range of substances, including fertilizers, pesticides, herbicides, and plant growth regulators [67].

It is important to note that while these substances offer significant benefits in terms of crop productivity and protection, their use should be done judiciously and in accordance with recommended practices and regulations, since their indiscriminate use can lead to ecosystem degradation [68]. Appropriated application techniques, dosage, timing, and safety precautions are essential to minimize potential environmental and human health risks associated with agrochemical use. Weeds and insects are significant biotic factors that negatively impact agriculture by diminishing crop yield, production, and efficiency. Consequently, the widespread use of herbicides and insecticides is employed to mitigate these issues and achieve increased production by controlling or reducing pest populations [68]. In this scenario, agricultural nanoadditives have emerged in recent years [69]. These nanoadditives use the principle of high ratio between surface area and volume of nanostructured materials to potentialize microbicidal activity against pathogens, as well as to function as a nutrient, supplementing plant growth, being that these particles are more easily absorbed by the plant organism [70, 71].

Among the various types of metallic nanoparticles, silver-, copper-, and zinc-based particles are preferentially used as antimicrobial agents. Of these, silver is the [72] one that presents the highest cost in precursor materials and in their preparation. Meanwhile, both particles based on copper and zinc are simpler and cheaper to prepare and are already widely applied in tests with plant organisms [73]. In addition, copper is an essential micronutrient in many organisms, being an integral part of many proteins and metalloenzymes, playing a significant role in plant health and nutrition [73]. Due to the differentiated properties of materials at the nanometer scale, copper nanoparticles (Cu NPs) present greater absorption and efficiency than when compared to their other forms already used in the market [70, 73].

Similar to copper, zinc is also a critical micronutrient in both animals and plants, and it is necessary for the structure and function of a wide range of macromolecules, including hundreds of enzymes. Zinc is the only metal to be involved in all six classes of enzymes: oxide-reductases, transferases, hydrolases, lyases, isomerases, and ligases. Zinc ions exist primarily as complexes with proteins and nucleic acids and participate in all aspects of intermediate metabolism. In addition, zinc is one of the most rapidly depleted elements of soils [74].

One of the concerns related to the utilization of nanoadditives revolves around the environmental impact caused by the production process of these materials [72, 75–77]. Chemical routes are still the most prevalent in the literature for larger-scale productions, and only in recent years green routes are being prospected [78–83]. It is also worth mentioning that part of these green routes uses relatively noble parts of plants, such as starch, tea leaves, among others, using food supplies to produce particles [81, 82].

Recently, it has been reported in the literature the possibility of applying bioprocesses for the manufacture of these materials [84–88]. Bioprocesses are highly viable since they demand low costs for industrial utilities and can present high transformation power through the correct selection of transforming organisms [85]. In addition, these organisms allow the use of inputs of low added value as their metabolites, being several organisms specialized in their biome of origin in the degradation of low-quality residues, such as lignocellulolytic residues.

Still, in the literature, studies highlight the importance of bioprocesses in the production of biogenic nanoparticles, which have greater interactions with organisms [84]. This fact comes from the natural coating performed during its synthesis, and the particle is covered with several biomolecules contained in the reaction medium. In addition, biogenic nanoparticles are produced without the generation of coproducts, as in many cases from chemical routes.

However, it is necessary to emphasize that given their direct and intentional application in the environment, nano-agrochemicals can be considered particularly critical in terms of possible environmental impact, as they represent the only diffuse and intentional source of nanoparticles in the environment [89]. Precisely because they are involved in several biological processes, the metals have the ability to activate cell death pathways when in high concentrations [90–92]. It is worth mentioning that these nanomaterials also have the capacity for human toxicity and ecotoxicity, if at high concentrations. Therefore, there is growing concern regarding the indiscriminate use of nanoparticles, as often classical risk assessment tools can fail due to the lack of information about the life cycle of these materials [89]. In addition, the accumulation of nanoparticles in the soil results in their greater absorption through the roots of the plants, showing toxic effects and inhibiting the growth of the applied cultivars [72, 93].

To solve this problem, we can use substances that fix the nanoparticles, preventing their distribution in the soil of application. One of the most used scaffoldings for this purpose is activated carbon. This material has high surface area, low cost, and acidic sites. All these characteristics make this material widely suitable for the fixation of metal oxide particles, acting as a mechanism of slow and controlled release of the applied particles. This type of carbonaceous material can still be produced from several different sources of carbon, presenting already in the literature relevant results for several residual biomasses of processes. It is now referred to the activated carbon produced using biochar [94–97].

López-Vargas et al. [98] developed a study about the foliar application of copper nanoparticles (Cu NPs) in the production of tomatoes, using different concentrations, as a result, they obtained that the application of Cu NPs increased the firmness of the tomato fruits, consequently increasing the shelf life of the fruits, in addition to inducing the accumulation of bioactive compounds such as vitamin *C*, *lycopene*, total phenols, and flavonoids in tomato fruits.

Other studies in the literature showed benefits with the use of copper nanoparticles, such as the increase in photosynthesis and stomatal conductance in the cultivation of peppers (*Capsicum annuum*) [99], and in the case of exposure to copper oxide nanoparticles (CuO NPs), an increase in the nutrient quality in chive (*Allium fistulosum*) cultivation [100] and increased chlorophyll photosynthesis and antioxidant enzyme activity in mustard (*Brassica juncea*) [101]. Nowadays, the application of microbial seed coating processes for seed inoculation is also proposed, and studies show promising results [102–107].

In the light of all the discussed topics in this chapter, an interesting prospect of circular economic has arrived. By bioremediating metal-rich effluents, a metal containing biomass is generated [8]. The metal can then be extracted and transformed into nanoparticles [58], and the remaining biomass can be turned into biochar [95]. By impregnating the biochar with metal, this composite material can then be used as a plant growth promoter and bacterial inoculation support to use in the phytoremediation of metal-contaminated soils [97, 108].

6. Conclusions

Bioremediation is a broad and very promising approach regarding metal effluents. This chapter focused in to report information generated between the years 2003 and 2023, bringing some robust and applicable technological solutions for not only removing metals, but also looking to turn them into a valuable coproduct for new commercial segments to industries. Many techniques for metal bioremediation, applying microorganisms and/or their byproducts, were brought together. Overall, copper effluents could emerge as a promising candidate coproduct for environmental and sustainable reuse, applying circular economy approaches.

Acknowledgements

The authors would like to thank Termomecanica S.A., FAPESP (Process: 2022/04262-9), CIM-EMBRAPII, CNPQ, and UNIFESP for financial support and human resources. This study was also financed by the Coordenação de Aperfeiçoamento de Pessoal de Nível Superior—Brasil (CAPES)—Finance Code 001.

Author details

Suzan P. Vasconcellos[1*], André Paganotti[1], Vitor G. Vital[1], Lidiane M. Santos Lima[1], Giovanna S.M. Paiva[1], L. Furlaneto de Lima[1], Enrique Moreira[1], Leticia O. Sousa[1], Guilherme G. Guerini[1], Vinicius T. Santos[2], Flavia G. Lobo[2], Márcio R. Silva[2], Diogo S. Pellosi[1*] and Ricardo A.G. Silva[1*]

1 Instituto de Ciências Ambientais, Químicas e Farmacêuticas—ICAQF, Universidade Federal de São Paulo—UNIFESP, Diadema, SP, Brazil

2 Department of Research and Development, Termomecanica São Paulo S.A., São Bernardo do Campo, SP, Brazil

*Address all correspondence to: suzan.pantaroto@unifesp.br; diogo.pellosi@unifesp.br and galdino.ricardo@unifesp.br

References

[1] Gregory CE. The importance of mineral production for any nation. In: Gregory CE, editor. A Concise History of Mining. CRC Press; 2021;**1**:165-168. DOI: 10.1201/9781003211327

[2] Vidal O, Rostom F, François C, Giraud G. Global trends in metal consumption and supply: The raw material-Energy Nexus. Elements. 2017;**13**:319-324. DOI: 10.2138/gselements.13.5.319

[3] Volesky B. Detoxification of metal-bearing effluents: Biosorption for the next century. Hydrometallurgy. 2001;**59**:203-216. DOI: 10.1016/S0304-386X(00)00160-2

[4] Singh A, Pal DB, Mohammad A, Alhazmi A, Haque S, Yoon T, et al. Biological remediation technologies for dyes and heavy metals in wastewater treatment: New insight. Bioresource Technology. 2022;**343**:126154. DOI: 10.1016/j.biortech.2021.126154

[5] Rafique M, Hajra S, Tahir MB, Gillani SSA, Irshad M. A review on sources of heavy metals, their toxicity and removal technique using physico-chemical processes from wastewater. Environmental Science and Pollution Research. 2022;**29**:16772-16781. DOI: 10.1007/s11356-022-18638-9

[6] Patel AK, Singhania RR, Albarico FPJB, Pandey A, Chen C-W, Dong C-D. Organic wastes bioremediation and its changing prospects. Science of the Total Environment. 2022;**824**:153889. DOI: 10.1016/j.scitotenv.2022.153889

[7] Tambat VS, Patel AK, Chen C-W, Raj T, Chang J-S, Singhania RR, et al. A sustainable vanadium bioremediation strategy from aqueous media by two potential green microalgae. Environmental Pollution. 2023;**323**:121247. DOI: 10.1016/j.envpol.2023.121247

[8] Malik A. Metal bioremediation through growing cells. Environment International. 2004;**30**:261-278. DOI: 10.1016/j.envint.2003.08.001

[9] Dursun AY, Uslu G, Cuci Y, Aksu Z. Bioaccumulation of copper (II), Lead (II) and chromium (VI) by growing *Aspergillus niger*. Process Biochemistry. 2003;**38**:1647-1651. DOI: 10.1016/S0032-9592(02)00075-4

[10] Gadd GM. Microbial influence on metal mobility and application for bioremediation. Geoderma. 2004;**122**:109-119. DOI: 10.1016/j.geoderma.2004.01.002

[11] PubMed-Bioremediation. Available from: https://pubmed.ncbi.nlm.nih.gov/?term=metal+bioremediation&sort=pubdate [Accessed: April 25, 2023]

[12] Akpor OB. Heavy metal pollutants in wastewater effluents: Sources, effects and remediation. Advances in Bioscience and Bioengineering. 2014;**2**:37. DOI: 10.11648/j.abb.20140204.11

[13] Brar SK, Verma M, Surampalli RY, Misra K, Tyagi RD, Meunier N, et al. Bioremediation of hazardous wastes—A review. Practice Periodical of Hazardous, Toxic, and Radioactive Waste Management. 2006;**10**:59-72. DOI: 10.1061/(ASCE)1090-025X(2006)10:2(59)

[14] Gavrilescu M. Removal of heavy metals from the environment by biosorption. Engineering in Life

Sciences. 2004;**4**:219-232. DOI: 10.1002/elsc.200420026

[15] Shah K, Nongkynrih JM. Metal hyperaccumulation and bioremediation. Biologia Plantarum. 2007;**51**:618-634

[16] Rehman A, Shakoori FR, Shakoori AR. Heavy metal resistant *Distigma Proteus* (Euglenophyta) isolated from industrial effluents and its possible role in bioremediation of contaminated wastewaters. World Journal of Microbiology and Biotechnology. 2007;**23**:753-758. DOI: 10.1007/s11274-006-9291-5

[17] Rehman A, Farooq H, Shakoori AR. Copper tolerant yeast, Candida tropicalis, isolated from industrial effluents: Its potential use in wastewater treatment. Pakistan Journal of Zoology. 2007;**39**:405-412

[18] Pérez-Rama M. Cadmium removal by living cells of the marine microalga Tetraselmis Suecica. Bioresource Technology. 2002;**84**:265-270. DOI: 10.1016/S0960-8524(02)00045-7

[19] Jiménez-Rodríguez AM, Durán-Barrantes MM, Borja R, Sánchez E, Colmenarejo MF, Raposo F. Heavy metals removal from acid mine drainage water using biogenic hydrogen sulphide and effluent from anaerobic treatment: Effect of pH. Journal of Hazardous Materials. 2009;**165**:759-765. DOI: 10.1016/j.jhazmat.2008.10.053

[20] MacHado MD, Soares HMVM, Soares EV. Removal of chromium, copper, and nickel from an electroplating effluent using a flocculent Brewer's yeast strain of *Saccharomyces cerevisiae*. Water, Air, and Soil Pollution. 2010;**212**:199-204. DOI: 10.1007/s11270-010-0332-1

[21] Kosolapov DB, Kuschk P, Vainshtein MB, Vatsourina AV, Wießner A, Kästner M, et al. Microbial processes of heavy metal removal from carbon-deficient effluents in constructed wetlands. Engineering in Life Sciences. 2004;**4**:403-411. DOI: 10.1002/elsc.200420048

[22] Stasinakis AS, Thomaidis NS. Fate and biotransformation of metal and metalloid species in biological wastewater treatment processes. Critical Reviews in Environmental Science and Technology. 2010;**40**:307-364. DOI: 10.1080/10643380802339026

[23] Bhattacharya A, Gupta A. Evaluation of *Acinetobacter* sp. B9 for Cr (VI) resistance and detoxification with potential application in bioremediation of heavy-metals-rich industrial wastewater. Environmental Science and Pollution Research. 2013;**20**:6628-6637. DOI: 10.1007/s11356-013-1728-4

[24] Giovanella P, Vieira GAL, Ramos Otero IV, Pais Pellizzer E, de Jesus Fontes B, Sette LD. Metal and organic pollutants bioremediation by extremophile microorganisms. Journal of Hazardous Materials. 2020;**382**:121024. DOI: 10.1016/j.jhazmat.2019.121024

[25] Zhao R, Wang B, Cai QT, Li XX, Liu M, Hu D, et al. Bioremediation of hexavalent chromium pollution by *Sporosarcina saromensis* M52 isolated from offshore sediments in Xiamen, China. Biomedical and Environmental Sciences. 2016;**29**:127-136. DOI: 10.3967/bes2016.014

[26] Medfu Tarekegn M, Zewdu Salilih F, Ishetu AI. Microbes used as a tool for bioremediation of heavy metal from the environment. Cogent Food & Agriculture. 2020;**6**:1783174. DOI: 10.1080/23311932.2020.1783174

[27] Chaturvedi MK. Studies on chromate removal by chromium-resistant

Bacillus sp. isolated from tannery effluent. Journal of environmental protection (Irvine, Calif). 2011;**2**:76-82. DOI: 10.4236/jep.2011.21008

[28] Bharagava RN, Mishra S. Hexavalent chromium reduction potential of *Cellulosimicrobium* sp. isolated from common effluent treatment plant of tannery industries. Ecotoxicology and Environmental Safety. 2018;**147**:102-109. DOI: 10.1016/j.ecoenv.2017.08.040

[29] Kumar R, Bhatia D, Singh R, Rani S, Bishnoi NR. Sorption of heavy metals from electroplating effluent using immobilized biomass *Trichoderma viride* in a continuous packed-bed column. International Biodeterioration & Biodegradation. 2011;**65**:1133-1139. DOI: 10.1016/j.ibiod.2011.09.003

[30] Kapahi M, Sachdeva S. Bioremediation options for heavy metal pollution. Journal of Health and Pollution. 2019;**9**:191203. DOI: 10.5696/2156-9614-9.24.191203

[31] Kang C-H, Oh SJ, Shin Y, Han S-H, Nam I-H, So J-S. Bioremediation of lead by Ureolytic bacteria isolated from soil at abandoned metal mines in South Korea. Ecological Engineering. 2015;**74**:402-407. DOI: 10.1016/j.ecoleng.2014.10.009

[32] Kang C-H, Kwon Y-J, So J-S. Bioremediation of heavy metals by using bacterial mixtures. Ecological Engineering. 2016;**89**:64-69. DOI: 10.1016/j.ecoleng.2016.01.023

[33] Guo H, Luo S, Chen L, Xiao X, Xi Q, Wei W, et al. Bioremediation of heavy metals by growing hyperaccumulaor endophytic bacterium *Bacillus* sp. L14. Bioresource Technology. 2010;**101**:8599-8605. DOI: 10.1016/j.biortech.2010.06.085

[34] Dixit R, Wasiullah, Malaviya D, Pandiyan K, Singh U, Sahu A, et al. Bioremediation of heavy metals from soil and aquatic environment: An overview of principles and criteria of fundamental processes. Sustainability. 2015;**7**:2189-2212. DOI: 10.3390/su7022189

[35] Chibuike GU, Obiora SC. Heavy metal polluted soils: Effect on plants and bioremediation methods. Applied and Environmental Soil Science. 2014;**2014**:1-12. DOI: 10.1155/2014/752708

[36] Islam MS, Kormoker T, Idris AM, Proshad R, Kabir MH, Ustaoğlu F. Plant-microbe-metal interactions for heavy metal bioremediation: A review. Crop & Pasture Science. 2021;**73**:181-201. DOI: 10.1071/CP21322

[37] Leong YK, Chang J-S. Bioremediation of heavy metals using microalgae: Recent advances and mechanisms. Bioresource Technology. 2020;**303**:122886. DOI: 10.1016/j.biortech.2020.122886

[38] Ameen FA, Hamdan AM, El-Naggar MY. Assessment of the heavy metal bioremediation efficiency of the novel marine lactic acid bacterium, Lactobacillus plantarum MF042018. Scientific Reports. 2020;**10**:314. DOI: 10.1038/s41598-019-57210-3

[39] Oziegbe O, Oluduro AO, Oziegbe EJ, Ahuekwe EF, Olorunsola SJ. Assessment of heavy metal bioremediation potential of bacterial isolates from landfill soils. Saudi Journal of Biological Sciences. 2021;**28**:3948-3956. DOI: 10.1016/j.sjbs.2021.03.072

[40] Kalaimurugan D, Balamuralikrishnan B, Durairaj K, Vasudhevan P, Shivakumar MS, Kaul T, et al. Isolation and characterization of heavy-metal-resistant bacteria and their applications in environmental bioremediation. International journal of

Environmental Science and Technology. 2020;**17**:1455-1462. DOI: 10.1007/s13762-019-02563-5

[41] Cui J, Xie Y, Sun T, Chen L, Zhang W. Deciphering and engineering photosynthetic cyanobacteria for heavy metal bioremediation. Science of the Total Environment. 2021;**761**:144111. DOI: 10.1016/j.scitotenv.2020.144111

[42] Chakdar H, Thapa S, Srivastava A, Shukla P. Genomic and proteomic insights into the heavy metal bioremediation by cyanobacteria. Journal of Hazardous Materials. 2022;**424**:127609. DOI: 10.1016/j.jhazmat.2021.127609

[43] Amin M, Tahir F, Ashfaq H, Akbar I, Razzaque N, Haider MN, et al. Decontamination of industrial wastewater using microalgae integrated with biotransformation of the biomass to green products. Energy Nexus. 2022;**6**:100089. DOI: 10.1016/j.nexus.2022.100089

[44] Sayqal A, Ahmed OB. Advances in heavy metal bioremediation: An overview. Applied Bionics and Biomechanics. 2021;**2021**:1-8. DOI: 10.1155/2021/1609149

[45] Li S, Hua T, Li F, Zhou Q. Bio-electro-Fenton systems for sustainable wastewater treatment: Mechanisms, novel configurations, recent advances, LCA and challenges. An updated review. Journal of Chemical Technology & Biotechnology. 2020;**95**:2083-2097. DOI: 10.1002/jctb.6332

[46] Li S, Hua T, Yuan C-S, Li B, Zhu X, Li F. Degradation pathways, microbial community and electricity properties analysis of antibiotic sulfamethoxazole by bio-electro-Fenton system. Bioresource Technology. 2020;**298**:122501. DOI: 10.1016/j.biortech.2019.122501

[47] Alharbi OML, Basheer AA, Khattab RA, Ali I. Health and environmental effects of persistent organic pollutants. Journal of Molecular Liquids. 2018;**263**:442-453. DOI: 10.1016/j.molliq.2018.05.029

[48] Zou R, Angelidaki I, Yang X, Tang K, Andersen HR, Zhang Y. Degradation of pharmaceuticals from wastewater in a 20-L continuous flow bio-electro-Fenton (BEF) system. Science of the Total Environment. 2020;**727**:138684. DOI: 10.1016/j.scitotenv.2020.138684

[49] Brillas E, Sirés I, Oturan MA. Electro-Fenton process and related electrochemical technologies based on Fenton's reaction chemistry. Chemical Reviews. 2009;**109**:6570-6631. DOI: 10.1021/cr900136g

[50] Teng X, Li J, Wang Z, Wei Z, Chen C, Du K, et al. Performance and mechanism of methylene blue degradation by an electrochemical process. RSC Advances. 2020;**10**:24712-24720. DOI: 10.1039/D0RA03963B

[51] Fenton HJH. LXXIII.—Oxidation of tartaric acid in presence of iron. Journal of the Chemical Society, Transactions. 1894;**65**:899-910. DOI: 10.1039/CT8946500899

[52] Tomat R, Vecchi E. Electrocatalytic production of OH radicals and their oxidative addition to benzene. Journal of Applied Electrochemistry. 1971;**1**:185-188. DOI: 10.1007/BF00616941

[53] Ruppert G, Bauer R, Heisler G. The photo-Fenton reaction — An effective photochemical wastewater treatment process. Journal of Photochemistry and Photobiology A: Chemistry. 1993;**73**:75-78. DOI: 10.1016/1010-6030(93)80035-8

[54] Ghatge S, Yang Y, Ko Y, Yoon Y, Ahn J-H, Kim JJ, et al. Degradation of sulfonated polyethylene by a bio-photo-Fenton approach using glucose oxidase immobilized on titanium dioxide. Journal of Hazardous Materials. 2022;**423**:127067. DOI: 10.1016/j.jhazmat.2021.127067

[55] Lastre-Acosta AM, Vicente R, Mora M, Jáuregui-Haza UJ, Arques A, Teixeira ACSC. Photo-Fenton reaction at mildly acidic conditions: Assessing the effect of bio-organic substances of different origin and characteristics through experimental design. Journal of Environmental Science and Health, Part A. 2019;**54**:711-720. DOI: 10.1080/10934529.2019.1585721

[56] Colombo A, Módenes AN, Góes Trigueros DE, Giordani da Costa SI, Borba FH, Espinoza-Quiñones FR. Treatment of sanitary landfill leachate by the combination of photo-Fenton and biological processes. Journal of Cleaner Production. 2019;**214**:145-153. DOI: 10.1016/j.jclepro.2018.12.310

[57] Sirtori C, Zapata A, Oller I, Gernjak W, Agüera A, Malato S. Decontamination industrial pharmaceutical wastewater by combining solar photo-Fenton and biological treatment. Water Research. 2009;**43**:661-668. DOI: 10.1016/j.watres.2008.11.013

[58] Goswami RK, Agrawal K, Shah MP, Verma P. Bioremediation of heavy metals from wastewater: A current perspective on microalgae-based future. Letters in Applied Microbiology. 2022;**75**:701-717. DOI: 10.1111/lam.13564

[59] Din MI, Rehan R. Synthesis, characterization, and applications of copper nanoparticles. Analytical Letters. 2017;**50**:50-62. DOI: 10.1080/00032719.2016.1172081

[60] Jiang J, Pi J, Cai J. The advancing of zinc oxide nanoparticles for biomedical applications. Bioinorganic Chemistry and Applications. 2018;**2018**:1-18. DOI: 10.1155/2018/1062562

[61] Chandra S, Kumar A, Tomar PK. Synthesis of Ni nanoparticles and their characterizations. Journal of Saudi Chemical Society. 2014;**18**:437-442. DOI: 10.1016/j.jscs.2011.09.008

[62] Patil S, Chandrasekaran R. Biogenic nanoparticles: A comprehensive perspective in synthesis, characterization, application and its challenges. Journal of Genetic Engineering and Biotechnology. 2020;**18**:67. DOI: 10.1186/s43141-020-00081-3

[63] Mughal B, Zaidi SZJ, Zhang X, Hassan SU. Biogenic nanoparticles: Synthesis, characterisation and applications. Applied Sciences. 2021;**11**:2598. DOI: 10.3390/app11062598

[64] Muthukrishnan L. Nanotechnology for cleaner leather production: A review. Environmental Chemistry Letters. 2021;**19**:2527-2549. DOI: 10.1007/s10311-020-01172-w

[65] Roduner E. Size matters: Why nanomaterials are different. Chemical Society Reviews. 2006;**35**:583. DOI: 10.1039/b502142c

[66] Kumari P, Alam M, Siddiqi WA. Usage of nanoparticles as adsorbents for waste water treatment: An emerging trend. Sustainable Materials and Technologies. 2019;**22**:e00128. DOI: 10.1016/j.susmat.2019.e00128

[67] Siviter H, Bailes EJ, Martin CD, Oliver TR, Koricheva J, Leadbeater E, et al. Agrochemicals interact synergistically to increase bee mortality. Nature.

2021;**596**:389-392. DOI: 10.1038/s41586-021-03787-7

[68] Meena R, Kumar S, Datta R, Lal R, Vijayakumar V, Brtnicky M, et al. Impact of agrochemicals on soil microbiota and management: A review. Land (Basel). 2020;**9**:34. DOI: 10.3390/land9020034

[69] Ghorbanpour, M, Bhargava, P, Varma, A, Choudhary, DK, editors. Biogenic Nano-Particles and their Use in Agro-Ecosystems. Singapore: Springer; 2020. ISBN 978-981-15-2984-9

[70] Bonner, JT. Why Size Matters. Princeton: Princeton University Press; 2007. ISBN 9781400837557

[71] Siddiqi KS, Husen A, Rao RAK. A review on biosynthesis of silver nanoparticles and their biocidal properties. Journal of Nanobiotechnology. 2018;**16**:14. DOI: 10.1186/s12951-018-0334-5

[72] Chhipa H. Chapter 6 - Applications of Nanotechnology in Agriculture. Methods in Microbiology. 2019;**46**:115-142

[73] Rai M, Ingle AP, Pandit R, Paralikar P, Shende S, Gupta I, et al. Copper and copper nanoparticles: Role in management of insect-pests and pathogenic microbes. Nanotechnology Reviews. 2018;**7**:303-315. DOI: 10.1515/ntrev-2018-0031

[74] Yuvaraj M, Surbramanian KS. Fabrication of zinc nano fertilizer on growth parameter of rice. BioScience Trends. 2014;7:2564-2565

[75] de Albuquerque Brocchi E, de Siqueira RNC, Motta MS, Moura FJ, Solórzano-Naranjo IG. Reduction reactions applied for synthesizing different nano-structured materials. Materials Chemistry and Physics. 2013;**140**:273-283. DOI: 10.1016/j.matchemphys.2013.03.034

[76] Ghosh Chaudhuri R, Paria S. Core/shell nanoparticles: Classes, properties, synthesis mechanisms, characterization, and applications. Chemical Reviews. 2012;**112**:2373-2433. DOI: 10.1021/cr100449n

[77] Shang Y, Hasan MK, Ahammed GJ, Li M, Yin H, Zhou J. Applications of nanotechnology in plant growth and crop protection: A review. Molecules. 2019;**24**:2558. DOI: 10.3390/molecules24142558

[78] Albrecht MA, Evans CW, Raston CL. Green chemistry and the health implications of nanoparticles. Green Chemistry. 2006;**8**:417-432. DOI: 10.1039/b517131h

[79] Dinda G, Halder D, Vazquez C, Lopez-Quintela M, Mitra A. Green synthesis of copper nanoparticles and their antibacterial property. Journal of Surface Science and Technology. 2015;**31**:117-122

[80] Thema FT, Manikandan E, Dhlamini MS, Maaza M. Green synthesis of ZnO nanoparticles via Agathosma betulina natural extract. Materials Letters. 2015;**161**:124-127. DOI: 10.1016/j.matlet.2015.08.052

[81] Umer A, Naveed S, Ramzan N, Rafique MS, Imran M. A green method for the synthesis of copper nanoparticles using L-ascorbic acid. Matéria (Rio de Janeiro). 2014;**19**:197-203. DOI: 10.1590/S1517-70762014000300002

[82] Rolim WR, Pelegrino MT, de Araújo Lima B, Ferraz LS, Costa FN, Bernardes JS, et al. Green tea extract mediated biogenic synthesis of silver nanoparticles: Characterization, cytotoxicity evaluation and antibacterial

activity. Applied Surface Science. 2019;**463**:66-74. DOI: 10.1016/j.apsusc.2018.08.203

[83] Valodkar M, Modi S, Pal A, Thakore S. Synthesis and anti-bacterial activity of Cu, Ag and Cu–Ag alloy nanoparticles: A green approach. Materials Research Bulletin. 2011;**46**:384-389. DOI: 10.1016/j.materresbull.2010.12.001

[84] Kasana RC, Panwar NR, Kaul RK, Kumar P. Copper nanoparticles in agriculture: Biological synthesis and antimicrobial activity. Nanoscience in Food and Agriculture 3. Sustainable Agriculture Reviews. Cham: Springer; 2016;**23**:129-143. DOI: 10.1007/978-3-319-48009-1_5

[85] Bukhari SI, Hamed MM, Al-Agamy MH, Gazwi HSS, Radwan HH, Youssif AM. Biosynthesis of copper oxide nanoparticles using streptomyces MHM38 and its biological applications. Journal of Nanomaterials. 2021;**2021**:1-16. DOI: 10.1155/2021/6693302

[86] Cuevas R, Durán N, Diez MC, Tortella GR, Rubilar O. Extracellular biosynthesis of copper and copper oxide nanoparticles by *Stereum Hirsutum*, a native white-rot fungus from Chilean forests. Journal of Nanomaterials. 2015;**2015**:1-7. DOI: 10.1155/2015/789089

[87] Ottoni CA, Simões MF, Fernandes S, dos Santos JG, da Silva ES, de Souza RFB, et al. Screening of filamentous fungi for antimicrobial silver nanoparticles synthesis. AMB Express. 2017;**7**:31. DOI: 10.1186/s13568-017-0332-2

[88] Noor S, Shah Z, Javed A, Ali A, Hussain SB, Zafar S, et al. A fungal based synthesis method for copper nanoparticles with the determination of anticancer, antidiabetic and antibacterial activities. Journal of Microbiological Methods. 2020;**174**:105966. DOI: 10.1016/j.mimet.2020.105966

[89] Kah M. Nanopesticides and nanofertilizers: Emerging contaminants or opportunities for risk mitigation? Frontiers in Chemistry. 2015;**3**:64. DOI: 10.3389/fchem.2015.00064

[90] Jeevanandam J, Barhoum A, Chan YS, Dufresne A, Danquah MK. Review on nanoparticles and nanostructured materials: History, sources, toxicity and regulations. Beilstein Journal of Nanotechnology. 2018;**9**:1050-1074. DOI: 10.3762/bjnano.9.98

[91] Love SA, Maurer-Jones MA, Thompson JW, Lin Y-S, Haynes CL. Assessing nanoparticle toxicity. Annual Review of Analytical Chemistry. 2012;**5**:181-205. DOI: 10.1146/annurev-anchem-062011-143134

[92] Zoroddu MA, Medici S, Ledda A, Nurchi VM, Lachowicz JI, Peana M. Toxicity of nanoparticles. Current Medicinal Chemistry. 2014;**21**:3873-3853

[93] Shobha G, Moses V, Amanda S. Biological synthesis of copper nanoparticles and its impact-a review. International Journal of Pharmaceutical Science Invention. 2014;**3**:28-38

[94] Chen M, Wang D, Yang F, Xu X, Xu N, Cao X. Transport and retention of biochar nanoparticles in a Paddy soil under environmentally-relevant solution chemistry conditions. Environmental Pollution. 2017;**230**:540-549. DOI: 10.1016/j.envpol.2017.06.101

[95] Oleszczuk P, Ćwikła-Bundyra W, Bogusz A, Skwarek E, Ok YS. Characterization of nanoparticles of biochars from different biomass. Journal

of Analytical and Applied Pyrolysis. 2016;**121**:165-172. DOI: 10.1016/j.jaap.2016.07.017

[96] Liu J, Jiang J, Meng Y, Aihemaiti A, Xu Y, Xiang H, et al. Preparation, environmental application and Prospect of biochar-supported metal nanoparticles: A review. Journal of Hazardous Materials. 2020;**388**:122026. DOI: 10.1016/j.jhazmat.2020.122026

[97] Zhang K, Wang Y, Mao J, Chen B. Effects of biochar nanoparticles on seed germination and seedling growth. Environmental Pollution. 2020;**256**:113409. DOI: 10.1016/j.envpol.2019.113409

[98] López-Vargas E, Ortega-Ortíz H, Cadenas-Pliego G, de Alba Romenus K, Cabrera, de la Fuente M, Benavides-Mendoza A, et al. Foliar application of copper nanoparticles increases the fruit quality and the content of bioactive compounds in tomatoes. Applied Sciences. 2018;**8**:1020. DOI: 10.3390/app8071020

[99] Rawat S, Cota-Ruiz K, Dou H, Pullagurala VLR, Zuverza-Mena N, White JC, et al. Soil-weathered Cuo nanoparticles compromise foliar health and pigment production in spinach (Spinacia Oleracea). Environmental Science & Technology. 2021;**55**:13504-13512. DOI: 10.1021/acs.est.0c06548

[100] Wang Y, Deng C, Cota-Ruiz K, Peralta-Videa JR, Sun Y, Rawat S, et al. Improvement of nutrient elements and allicin content in green onion (*Allium fistulosum*) plants exposed to CuO nanoparticles. Science of the Total Environment. 2020;**725**:138387. DOI: 10.1016/j.scitotenv.2020.138387

[101] Faraz A, Faizan M, Hayat S, Alam P. Foliar application of copper oxide nanoparticles increases the photosynthetic efficiency and antioxidant activity in *Brassica juncea*. Journal of Food Quality. 2022;**2022**:10. Article ID 5535100. DOI: 10.1155/2022/5535100

[102] Rouphael Y, Colla G, Graziani G, Ritieni A, Cardarelli M, De Pascale S, et al. Antioxidant activity and mineral profile in two seed-propagated artichoke cultivars as affected by microbial inoculants and planting time. Food Chemistry. 2017;**234**:10-19. DOI: 10.1016/j.foodchem.2017.04.175

[103] Paravar A, Piri R, Balouchi H, Ma Y. Microbial seed coating: An attractive tool for sustainable agriculture. Biotechnology Reports. 2023;**37**:e00781

[104] Afzal M, Khan S, Iqbal S, Mirza MS, Khan QM. Inoculation method affects colonization and activity of Burkholderia phytofirmans PsJN during phytoremediation of diesel-contaminated soil. International Biodeterioration & Biodegradation. 2013;**85**:331-336. DOI: 10.1016/j.ibiod.2013.08.022

[105] Barnett SJ, Ballard RA, Franco CMM. Field assessment of microbial inoculants to control Rhizoctonia root rot on wheat. Biological Control. 2019;**132**:152-160. DOI: 10.1016/j.biocontrol.2019.02.019

[106] Seleiman MF, Ali S, Refay Y, Rizwan M, Alhammad BA, El-Hendawy SE. Chromium resistant microbes and melatonin reduced Cr uptake and toxicity, improved physio-biochemical traits and yield of wheat in contaminated soil. Chemosphere. 2020;**250**:126239. DOI: 10.1016/j.chemosphere.2020.126239

[107] Imdad, Ali Mahmood AAMAU, Nawaz Q, Badar-Uz-Zaman TS. Effect of inoculation methods of biozote-max

(Plant Growth Promoting Rhizobacteria-PGPR) on growth and yield of rice under naturally salt-affected soil. Research in Plant Biology. 2017;7:24-26. DOI: 10.25081/ripb.2017.v7.3602

[108] Yaashikaa PR, Kumar PS, Jeevanantham S, Saravanan R. A review on bioremediation approach for heavy metal detoxification and accumulation in plants. Environmental Pollution. 2022;**301**:119035. DOI: 10.1016/j.envpol.2022.119035

Chapter 6

Analysis of the Oxidation-Reduction Potential and Bacterial Population of *Acidithiobacillus ferrooxidans* during the Bioleaching Study of Sulfide Ores

Vladimir Arias-Arce, Daniel Lovera-Dávila, José J. Guerrero-Rojas, Fanny Blas-Rodriguez and Ismael Molina-Pereyra

Abstract

The analysis of the variables, bacterial population, and oxidation-reduction potential (ORP) during the bioleaching of sulfide ores by a bacterial strain of *Acidithiobacillus ferrooxidans*, isolated from acid mine effluent, aims at the solubilization of copper and the liberation of the gold present in an ore containing more than 80% sulfides. It was studied at different pulp densities (1, 2, and 6% - W/V) and with a 9 k medium at different ferrous sulfate concentrations (0, 3, 6, 9, 12, and 15 g/L), keeping temperature and pH constant. The tests were carried out in three consecutive stages, starting with inoculum, whose cell content was $7.05x10^7$ Cell/mL, then the strain with the highest population obtained in the previous stage was used, observing the variation in the periods of adaptation and growth. During the bioleaching of sulfide ores, in the first stage, the maximum bacterial population achieved was $4.75x10^7$ Cell/mL in 24 days with 6 g/L ferrous sulfate, in the second stage, the maximum population was $6.30x10^7$ Cell/mL without the addition of ferrous sulfate, and in the third stage, the bacterial population became $4.51x10^7$ Cell/mL. The exponential characteristic growth of the population started at approximately 13, 8, and 3 days, respectively in each stage.

Keywords: bacterial population, bioleaching, sulfide minerals, *Acidithiobacillus ferrooxidans*, redox potential

1. Introduction

The redox potential of a mineral solution or ore pulp is a measure of electron activity that can be influenced by the presence of pyrite in its natural status, considered electrochemically passive, a favorable condition for the galvanic effect with other sulfide compounds to be enhanced and the formation of elemental sulfur to be

IntechOpen

achieved [1]. Therefore, several works are carried out to identify the variables and parameters that allow the liberation of metallic and nonmetallic ions that are present in a mineral.

Specifically, the identification of the oxidation-reduction potential (ORP), which is a critical factor in the development of the inoculum and in the evolution of the oxidation of inorganic compounds, is determined with a platinum reference electrode and a hydrogen electrode connected to a potentiometer. It is quantified in volts, which represent the energy released by all components of the system in a fraction of time when a number of electrons move from one phase to another: namely, between the bioleaching substrate and the platinum electrode.

The biological oxidation of sulfide to elemental sulfur, sulfate anion, and other sulfur compounds and the reduction of oxygen in water are the main redox changes that occur in this process. The measured redox value of the medium in which a process takes place will be the result of the set of chemical reactions occurring in it.

Likewise, the thermodynamic relationship of the oxidation-reduction potential (ORP) represented by the potential (Eh) of the solution is known as the Nernst equation, nevertheless, in practice, the ORP value is determined mostly by ionic compounds with high current exchange density, in other words, the ability they have to exchange electrons on the surface of the platinum electrode; In this sense, several researchers have found that there are compounds that have a great aptitude to exchange their valence electrons on the platinum surface, such is the case of hydrogen sulfide, for which there is a linear relationship between the ORP measurement and the logarithm of the concentration of hydrogen sulfide in natural environments [2]. The process of dissolution of chalcopyrite with sulfur-oxidizing microorganisms and iron oxidizers depends mainly on the redox potential. Chalcopyrite was preferentially oxidized to polysulfide when the redox potential is approximately less than 350 mV with reference to Ag/AgCl electrode and at higher potential approximately between 350 and 480 mV, chalcopyrite was mainly transformed to Cu_2S, intermediate species, resulting in high dissolution rate and when the redox potential is higher than that of 480 mV, chalcopyrite was mostly oxidized to polysulfides, causing passivation of chalcopyrite [3].

The usefulness of ORP data may be questionable because the measuring probe is directly in contact with the sinus of the extracellular environment, which is totally different from the intracellular environment. A disadvantage of ORP is its strong dependence on pH. In this regard, decreases in potential of up to 33 mV have been reported with a one-unit increase in pH [4]. However, real-time potential monitoring offers many advantages [5].

It is known that in culture media, microorganisms show very different sensitivities to the oxidation-reduction potential. Therefore, it is believed that the redox potential is a very particular and important factor in each of the environments where the substrate probably determines the presence of a variety of microorganisms and their metabolic evolution [6].

By means of bioaugmentation processes under conventional bioleaching conditions, the growth of diverse bacteria was achieved, which contributed to improve copper dissolution, achieving the extraction of 90.2% after 24 days. In the final stage, the formation of a passivating layer of jarosite decreases the copper release rate; resulting that, the increase of iron-oxidizing cells negatively influencing the leaching of chalcopyrite [7].

It should be noted that sulfide minerals often coexist with several mineral species, which act synergistically during the leaching process, allowing the dissolution of certain elements. In the case of leaching of chalcopyrite and silver-bearing bornite [8], by thermophilic culture (50°C), 94.6% copper was extracted, quite superior to the results of separate leaching of chalcopyrite and bornite, silver was released to the solution, forming Ag_2S on the surface of chalcopyrite. In recent studies of bioprocesses, bioleaching is considered a process with multiple advantages such as low cost, environmental friendliness, simplicity of requirements, and suitability for the treatment of low-grade ores. In the analysis of commercial processes, the evolution of the problem is identified as the lack of definition of the parameters due to the synergic effect of microorganisms, the role of extracellular polymeric substances, the passivation phenomenon, the galvanic interaction between minerals, the mode of application, and the environment [9].

2. Background of bioprocesses

In recent years, the development of microbiological processes for the extraction of metals from ore bodies has generated much interest and the approach to biotechnological processes such as biooxidation, which are already applied in many parts of the world. These are the fundamental reasons for research and for providing incentives for new discoveries and are also likely to become the cause of the development of the mining and metallurgical sector [10, 11]. Bioleaching is a clean technology for processing complex and low-grade ores because of reduced energy and water consumption and low CO_2 and SO_2, emissions compared to pyrometallurgical processes [12].

In the bioleaching of copper sulfide ores using *At. ferrooxidans* and subsequent characterization of the residues by scanning electron microscopy (SEM) and X-ray diffraction (XRD), it has been identified that dissolution occurs in the following preferential order: bornite, pyritic chalcopyrite, covellite, and porphyry chalcopyrite; the latter as a surface layer can hinder the dissolution of other compounds and thus, the extraction of copper [13, 14]. Polysulfides, elemental sulfur, and insoluble sulfates are the main constituents that determine the redox potential [15].

It can be asserted that electron donor sources are abundant and diverse in nature and can be of anthropogenic, geological, biological, and inorganic materials. An important source of inorganic compounds is volcanic activity as reduced sulfur compounds and others. All compounds derived from the mining and agricultural industries, products from the burning of hydrocarbons, and other industrial activities release reduced sulfur compounds into the environment, which donate or receive electrons and thus energy through chemolithoautotrophic sulfoxidizing bacteria [16].

During the last three decades, the application of bioleaching for the treatment of sulfide ores has reached its industrialization, the sequential use of biooxidation —bioleaching—electrowinning, for the extraction of copper, uranium, gold, and zinc, providing satisfaction in the mining sector. In addition, in recent years, its application is being sought for the extraction of copper from refractory ores [17]. In the dissolution of chalcopyrite promoted with ferric ions, Hiroyoshi et al. [18] presented a two-stage dissolution model: first, the reduction of chalcopyrite to Cu_2S by ferrous ions in the presence of cupric ions. Second, the oxidation of Cu_2S to cupric ions and elemental sulfur. Reactions achieved at solution potentials below the predicted critical potential as a function of ferrous, ferric, and cupric ion concentrations.

In studies by Nazari et al. [19], it was observed that a ferric precipitate in the form of jarosite was produced at 50 g/L ferrous sulfate at an initial pH of 2.2 and a temperature of 32°C. The effects of ferric iron precipitation on other ions are important for *At. ferrooxidans* bacteria in the aqueous phase, that is, sulfate, phosphate, magnesium, and potassium ions. The results showed relatively similar patterns for potassium and ferric ions, and this could be explained by the coprecipitation of these ions as constituents of jarosite, increased at higher pH.

The copper extraction yield from thermophilic bioleaching of chalcopyrite depends on temperature, pH, and oxidation-reduction potential (ORP), as well as the activity of the thermophile used [20]. The copper extraction yields obtained with thermophilic microorganisms at different pH and temperature conditions, and with different initial amounts of Fe^{3+}, generate high biomass concentrations at an ORP close to a critical value (450 mV, with reference to the $Ag^0/AgCl$ electrode) and high copper extraction, attenuating at higher ORP values and causing Fe3+ precipitation as jarosite [21]. However, the dissolution of chalcopyrite might not be hindered even though large amounts of jarosite are produced and the free jarosite would be easily detached from the ore surface [22, 23].

Through tests with different electrochemical circuits [24] for the dissolution of chalcopyrite, it was identified that the increase in potential caused the formation of a CuS layer, hindering the dissolution speed of the electrode. The formation of CuS is concomitant with the formation of $Fe_2(SO_4)_3$ and the latter can act as a precursor to jarosite nucleation at potentials around 750 mV (referring to the Hg°/Hg_2SO_4 electrode). Concluding with the modeling of the experimental results. Also, Zhao et al. [20], through thermodynamic calculations and electrochemical measurements, established the conditions to determine the optimum potential in the leaching of chalcopyrite, having as main variables the temperature and the concentrations of Cu^{2+} and Fe^{2+}, managing to establish a model to predict the potential range, becoming inhibited due to the formation of jarosite, requiring the periodic addition of Cu^{2+} and Fe^{2+} ions to improve the bioleaching of chalcopyrite. In this line, Yang et al. [23], during the electrochemical oxidation of a chalcopyrite electrode at potentials between 550 and 630 Mv (having as reference an Ag/AgCl electrode), finds an anodic dissolution region with S_2^{2-} and covellite species and two very close passive regions coated with a thin sulfur-rich layer, which could be responsible for the passivation, concluding that chalcopyrite can passivate in the potential range of 748–828 mV with respect to the standard hydrogen electrospray (SHE). On the other hand, at an applied potential of 415 to 750 mV, a thin film of copper and iron sulfide was produced, exhibiting passivation properties and, at 1070 mV, the film formed dissolved and the rate of chalcopyrite dissolution was enhanced; when the potential continued to increase, CuS was formed and hindered chalcopyrite dissolution; finally, at 1400 mV, jarosite was formed, which hindered chalcopyrite dissolution [24]. Subsequently, to investigate the roles of dissolved oxygen (O_2), Fe^{3+} and Fe^{2+} and their interactions during the leaching of chalcopyrite in a basic culture medium at atmospheric pressure and 45°C, it was shown that Fe^{3+} significantly promoted the dissolution of chalcopyrite in the initial stage, then in the final stage it caused the passivation by the formation of jarosite due to the oxidation of Fe^{3+}, it was also tested adding oxygen, which caused an appropriate potential range between 380 and 480 mV (with respect to the Ag/AgCl electrode), eliminating the passivation species caused by polysulfides and favoring the formation of jarosite [25].

The hindrance or delay in the dissolution of minerals by bacterial action is due to the formation of a surface layer, a phenomenon known as passivation, which is the subject of debate. In leaching tests, in the presence and absence of mixed culture, it was found that the presence of jarosite and elemental sulfur in an abiotic experiment does not hinder copper dissolution and the dissolution curves do not represent signs of postdissolution but indicated hindered dissolution. In bioleaching and abiotic tests

DOI: http://dx.doi.org/10.5772/intechopen.111815

with chalcopyrite samples, it was identified that the common phases on the surface of the leached samples during different periods of time were elemental sulfur and iron oxyhydroxides, which were identified by XPS spectrometer as jarosite, being the cause of the difficulty in dissolution [26]. The kinetics of chalcopyrite dissolution is fast when the solution potential is lower than 648 mV (SHE), and it cannot be effectively leached when the solution potential is higher than 698 mV due to the production of polysulfides, elemental sulfur, and jarosite on the surface, reaching surface passivation; it is not possible to oxidize with Fe^{3+}, but it can be oxidized by stronger oxidants [8].

3. Experimental procedure

In leaching with *At. ferrooxidans*, the conditions at which various works have been carried out by several researchers, including Wang et al. [17], identify that at the temperature of 30 ± 1°C and pH of 2.0 achieve concentrations of $2.24x10^7$ Cell/mL and recovery of 45% copper after 75 days of leaching. Liu et al. [22], during the bioleaching of chalcopyrite at different times, identifies the presence of various species such as bornite, chalcocite, covellite, and their respective redox potentials and, finally, after 30 days of processing, identifies as iron species approximately 26% chalcopyrite, 10.2% bornite, and 74% jarosite.

Seeking to contribute to the knowledge of the mechanisms of bioleaching of sulfur minerals, in this study, the experimental design, implementation, and execution of the tests were carried out in the Laboratory of Biometallurgy of the School of Metallurgical Engineering of the Universidad Nacional Mayor de San Marcos (UNMSM) with the participation of teachers and students of the faculties of Chemistry and Chemical Engineering and Biological Sciences. Potential (Ev) measurements and bacterial population determinations were carried out at different concentrations of 9 k substrate and mineral substrate, maintaining constant pH, temperature, and agitation speed. The test medium consisted of sulfide mineral, 100 mL of 9 k solution, 10 ml of inoculum, pH of 1.8, average ambient temperature of 22°C, and constant agitation of 150 rpm. The analyses to determine the extraction of Cu and other elements were carried out by Atomic Absorption and Induced Plasma Spectrometry, in the laboratory of the School of Metallurgical Engineering and by third-party service, respectively. The bacterial population count was carried out at the Environmental Microbiology and Biotechnology laboratory of the School of Biological Sciences, UNMSM.

3.1 Mineral substrate as metabolic medium

An important factor of the ore under investigation is its nature; with the presence of diverse sulfides, being of interest the copper sulfides. The sulfide ore was milled at 94% -200 Mesh to facilitate its oxidation and the supply of nutrients required by the inoculated microorganism.

The mineralogical composition of the sulfide ore that forms part of the substrate in the bioleaching was identified, containing mainly: iron, sulfur, copper, gold, and silver. As well as high content of gangue with sulfide compounds that will increase the pH in the leach liquor, with the consequent inhibition and suppression of microbial activity [27]. The degree of leaching to be achieved will depend on the type of surface of the mineral substrate, as a decrease in particle size means an increase in specific surface area so that dissolution or oxidation yields can be obtained without any alteration of the total particle mass. A particle size of approximately 42 μm is considered optimal [28]. Additionally, the provision of a 9 k culture broth modified in its ferrous sulfate content favors the reactivity of the medium [29].

3.2 Isolation in solid and liquid media 9 k

First, colony morphology, cell morphology, ferrous ion oxidation, and tetrathionate oxidation were considered in the presumptive identification of the isolated bacteria. The colonies are then poured into solid medium (Petri plates) and allowed to gel by cooling. In total, 0.1 ml of bacterial strain from liquid culture was added. It is placed in an incubator at 28°C for periods of 5 to 10 days. The evaluation of growth was by direct observation from the fifth day based on the methodology of Hallberg et al. [30]. Reddish brown colonies were obtained due to the formation of ferric iron. The identified colony is reseeded in fresh liquid medium, thus achieving the enrichment of the strain and obtaining the intrinsic characteristics of *At. ferrooxidans*. Iron is used in both isolates because it is an essential micronutrient for bacteria and because of its oxidative and reductive properties, it acts as an electron transporter and as a cofactor for many enzymes. Subsequently, the pure strain was sent to the UNMSM Molecular Biology laboratory for identification and final characterization using the polymerase chain reaction (PCR) technique.

The Wizard Genomic DNA Purification Kit (Promega) protocol was used for chromosomal DNA extraction. We then proceeded to design the universal primers that amplify the 16S ribosomal RNA coding gene. DNA sequencing used the dye termination method and the Applied Biosystems 3730 system from Macrogen USA was used. And finally, molecular identification was carried out by comparing the 16S rRNA gene sequences (16S rDNA) of the native strains with those available in the databases using the program BlastN version 2.0. The 16S rRNA sequences were obtained from GenBank/EMBL/DDBJ databases, according to the percentage of similarity, the RecB1a isolate was identified as *At. ferrooxidans* in 98% [31].

3.3 Bacterial adaptation in the presence of metallic sulfides

Adaptation tests are carried out by several researchers seeking to obtain a bacterium or bacterial consortium capable of growing in media consisting of mineral sulfides and sulfides refractory to conventional mineral extraction processes. We cite some studies carried out on the adaptation in minerals with the presence of arsenic and silver sulfides, being these compounds inhibitors to bacterial development. The tests consisted of adapting the bacterial strain to different media with different amounts of ferrous sulfate and pyritic sulfide mineral [32].

Once the 9 k medium was identified, modifying its iron content and at a favorable pH, adaptation was sought in the presence of minerals containing about 70% sulfides of various mineral species such as chalcopyrite, pyrrhotite, arsenopyrite, and others. The evolution or progress of the adaptation phase is determined by the variation of the redox potential [33], determining that above 500 mV a marked bacterial growth occurs, then inhibition, and finally the decrease of the population; possibly, by the saturation of the medium and/or by nutrient depletion.

The objective of this adaptation stage in the application of a bioprocess such as bioleaching is to obtain a bacterial microorganisms capable of growing in sulfur media, without the addition of iron sulfate; in other words, to make the microorganisms feed themselves with the iron contained in the material subjected to the bioleaching process, this would help us to later carry out oxidation tests on refractory minerals containing high contents of pyrite, arsenopyrite and also, with arsenic and silver contents.

The isolated and identified bacterium (*At. ferrooxidans*) is subjected to media with mineral of approximately 80% sulfides, 10% arsenic, and 20% silicates, taking different amounts of mineral (1, 2, 3, and 4 g) and at different particle sizes. The basic conditions of the adaptation were: room temperature at an average of 20°C, pH 1.8, and shaker agitation at 150 rpm [34].

3.3.1 Adaptation to different quantities of ore

Four assays were carried out, with 1, 2, 3, and 4 grams of mineral pulverized at −200 mesh. In 250 mL Erlenmeyer flasks, 100 mL of 9 k medium and 10 mL of solution with *At. ferrooxidans* (pure culture) were added. The population and redox potential increased over time. The increase in potential occurs after a short latency period of approximately 10 to 12 hours. See **Figure 1**.

3.3.2 Adaptation to different ore particle sizes

Tests were also carried out with 100 ml of 9 k medium modified in its ferrous sulfate content and 10 mL of bacterial strain. Tests were carried out with the ore pulverized at −200 mesh and fractioned in three sizes: −200 + 325, −325 + 400, and − 400 mesh, taking 5 grams of each of the ore fractions. The ferrous sulfate content was limited to 11 g/L. The test with the mineral whose fraction corresponds to sizes smaller than 37 microns (−400 M) was of particular note. See **Figure 2**.

3.3.3 Prolonged adaptation

The test is carried out with the mineral pulverized at 80% - 400 mesh in 9 k liquid medium with 5.5 g/L of ferrous sulfate and 7.5 grams of the mineral. The adaptation

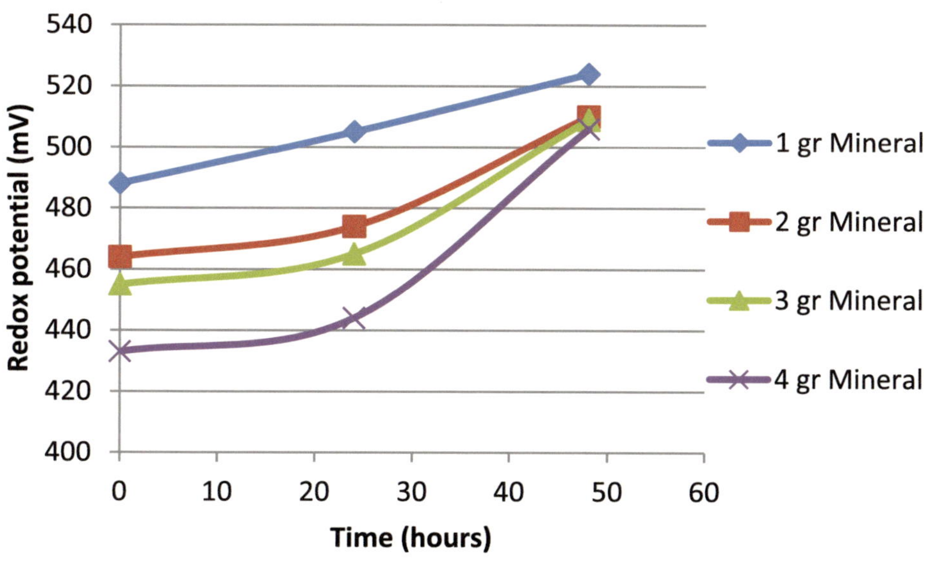

Figure 1.
Increase of the redox potential as a function of time. A greater increase in potential is achieved in the test with less mineral, consequently, less friction and less dissolution of its components, and less damage to the bacteria.

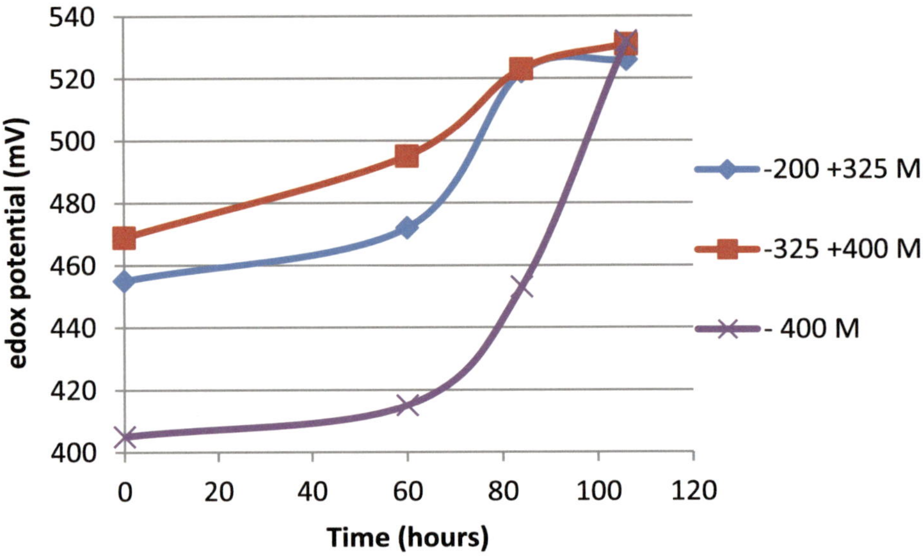

Figure 2.
Oxidation potential changes as a function of time with particle size fractions. A latency period of approximately 40 hours is observed in the smallest size fraction, followed by an exponential increase in redox potential.

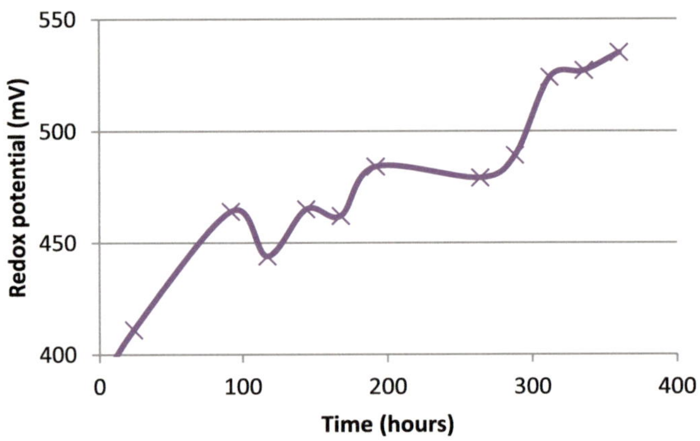

Figure 3.
Redox potential vs. adaptation time. The increase of the redox potential over a prolonged period represents the adaptation of the bacteria, the growth of its population, and consequently, the oxidation of the mineral.

is carried out for 380 hours, observing the increase of the potential with periods of inhibition. See **Figure 3**.

4. Bioleaching tests

The first tests performed corresponded to the chemical analysis of identification by elements, the results of which were subjected to theoretical analysis based on bibliographic information in order to define the operating parameters, as was done for the bioleaching tests of copper sulfide ores [28].

It is known that in bioleaching tests over a prolonged period of 60 days, the bacterial population is maintained around 550–590 mV [35]. Also, the evolution of the bacterial population shows increases in a certain period of time, for example, between 6 and 21 days of processing, the average bacterial population was 1.70x10^8 Cell/mL and 8.00x10^7 Cell/mL in the bioleaching of ore whose granulometries corresponded to −200 and − 325 Tyler mesh, respectively [33].

Bioleaching is performed in three consecutive stages with 1.0, 6.0, and 2.0% (W/V) of sulfide ore, the second and third stages using the best inoculum from the previous stage and under the conditions detailed in **Figure 4**.

A 9 k solution with different concentrations of ferrous sulfate is provided as nutrient substrate. During the tests, measurements of oxidation-reduction potential (ORP) and pH are taken; in addition, periodic sampling is carried out to determine the metals present, such as copper, iron, arsenic, and zinc.

4.1 First stage of bioleaching

The bioleaching solution had as main nutrient substrate medium 9 k at different concentrations of $FeSO_4.7H_2O$ between 0.0 to 15.0 g/L. For the assays, 500 mL Erlenmeyer flasks were used, where 3 g of mineral (1% W/V), 30 mL of bacterial strain of 7.05x10^7 Cell/mL, and 300 mL of 9 k solution containing: 3.0, 6.0, and 9.0 g/L of $FeSO_4.7H_2O$ were added. The pH was regulated to 1.8 by adding sulfuric acid solution. The process was carried out in an agitation platform (Orbit Shaker) at 150 rpm. According to research, the dose of ferrous ions for the bioleaching of sulfides such as pyrite and chalcopyrite differs depending on the characteristics of the mineral that contains them [36].

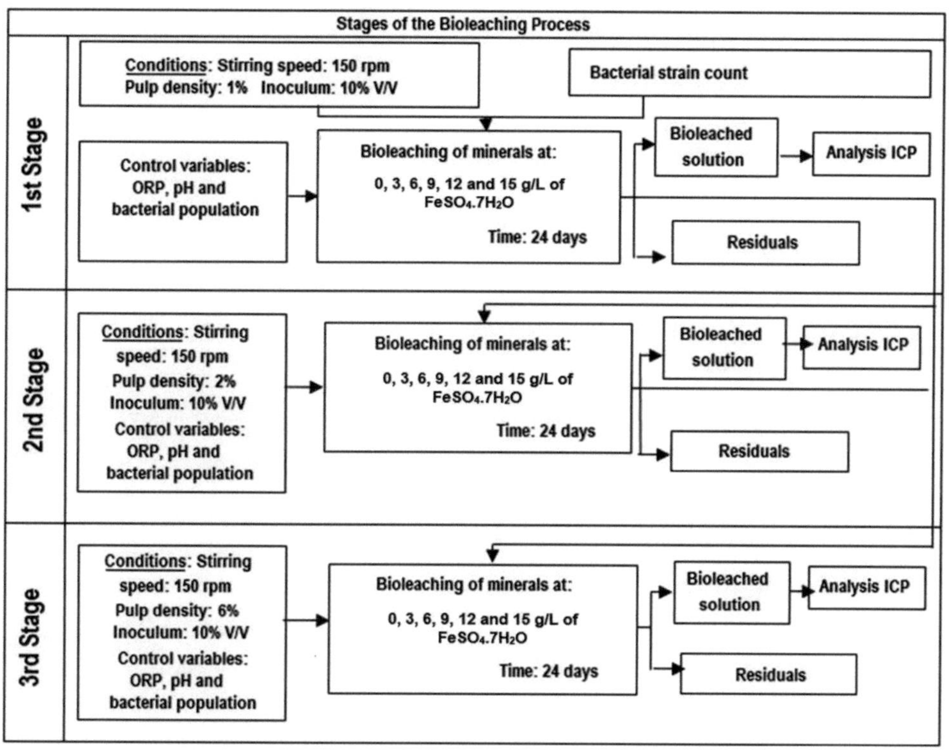

Figure 4.
Stage bioleaching design with 1, 2, and 6% solids.

The maximum copper recovery obtained in this stage was 72.64%, with 6.0 g/L of $FeSO_4.7H_2O$ and the minimum recovery was 38.96% with 15.0 g/L of $FeSO_4.7H_2O$. Approximately 20 days after the start of the process, it is observed that the copper dissolution is drastically reduced. See **Figure 5**.

4.1.1 Effect on bacterial population growth

Bacterial growth was identified as a function of the ferrous sulfate content provided, with an accelerated increase observed between approximately the 12th and 24th day, followed by a break, indicating the end of the bacterial population growth stage. The maximum population density reached was at 24 days with $4.75x10^7$ Cell/mL with 6.0 g/L of $FeSO_4.7H_2O$ reaching 67% of the inoculum, as shown in **Figure 6**.

The methodology used in bioleaching processes takes into account the adaptation stage, which is progressive for *A. ferrooxidans* bacteria in the presence of nutrient ions (Arias et al., 2015), where the reproduction of microorganisms is achieved; in parallel, the metal compounds in solution are increased [33, 37].

4.1.2 pH variation

The tests are started at pH 1.8 and at different concentrations of $FeSO_4.7H_2O$. During the first stage period, which lasts 19 days, the variation is observed and regulated. The pH varies from a minimum of 1.5 to a maximum of 2.2, on average 1.9. Finally, the trend is downward, probably due to the appearance of H^+ and the formation of sulfuric acid, which can be seen in **Figure 7**.

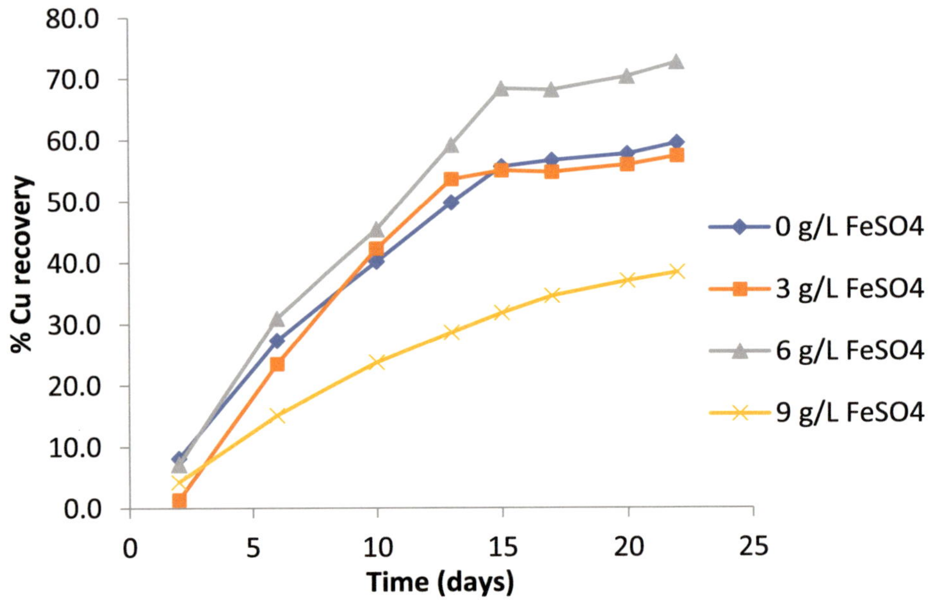

Figure 5.
Bioleaching in pulp containing 1% solids.

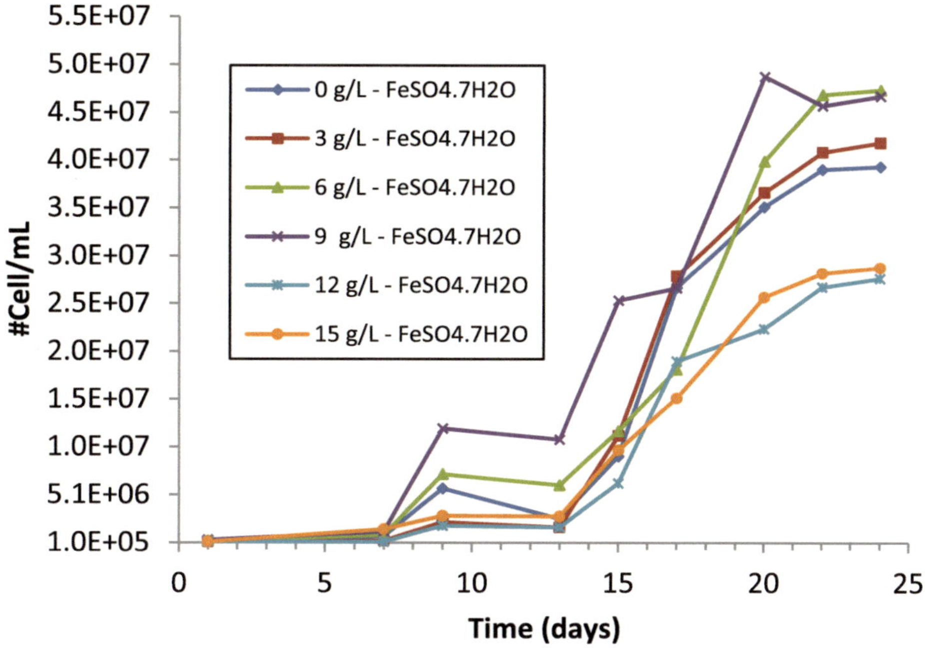

Figure 6.
Variation of the bacterial population during the first stage of bioleaching.

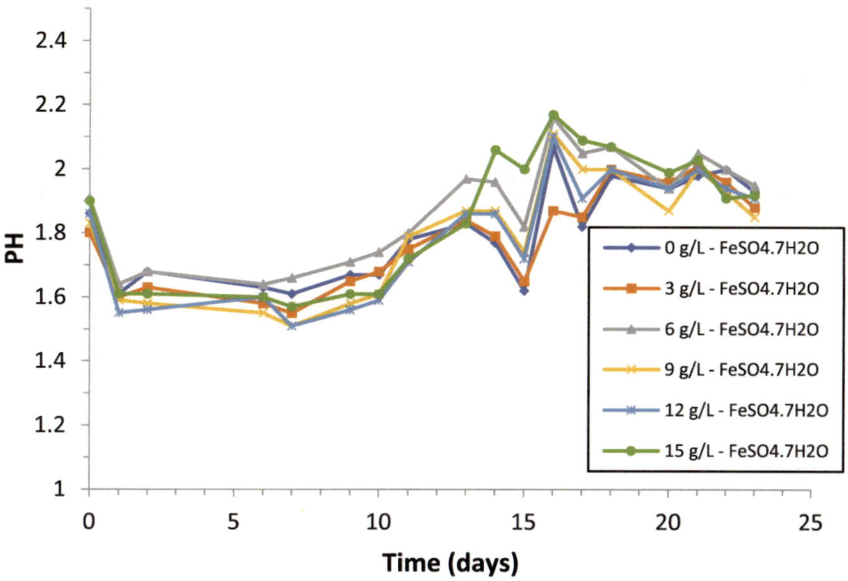

Figure 7.
pH variation during the first stage of bioleaching.

4.1.3 Measurement of oxidation: reduction potential

Figure 8 shows the ORP values for each test performed; in the first 6 days, the test containing no ferrous sulfate increases from 360 mV to 585 mV on approximately

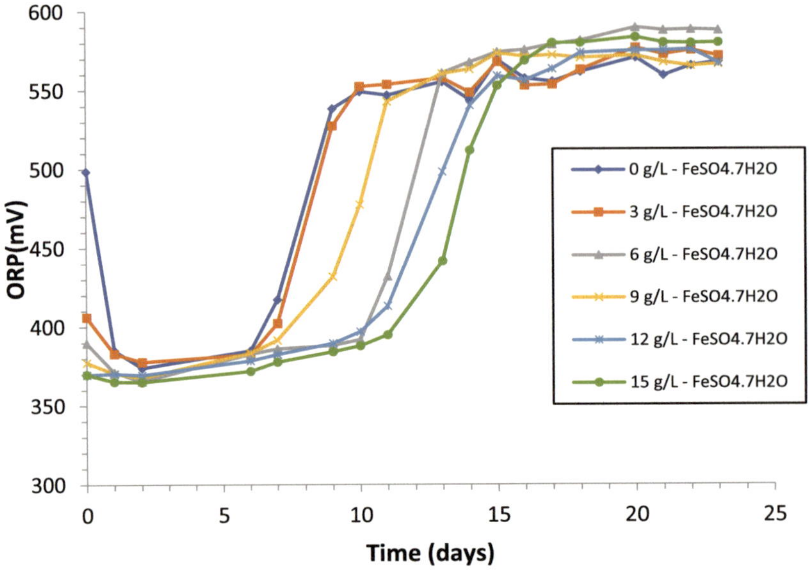

Figure 8.
Measurement of oxidation-reduction potential (ORP). First stage of the bioleaching process.

the tenth day. On the other hand, the sample containing 15.0 g/L of $FeSO_4.7H_2O$ achieves its increase after 10 days from the start of the test to approximately 560 mV. Subsequently, all samples are maintained at around 575 mV.

4.2 Second stage of bioleaching

Tests were performed in 500 mL Erlemeyer flasks, adding 18.0 g of mineral (6% W/V), 30 mL of bacterial strain (10% V/V), and 300 mL of 9 k Medium with 0.0, 2.0, 4.0, and 6.0 g/L of $FeSO_4.7H_2O$. The pH is regulated to 1.9 with sulfuric acid solution, the process was continued on a stirring platform at 150 rpm. Bacterial strain from the first stage is used. After 22 days of leaching, 89.38% copper extraction is achieved in a medium without the addition of ferrous sulfate. See **Figure 9**.

4.2.1 Effect on bacterial growth

Observed from the inoculation with the highest population strain obtained in the previous stage, whose concentration was $4.75x10^7$ Cell/mL.

It is observed that the adaptive and exponential phases show the same trend in all tests. The exponential growth phase occurs approximately between the 8th and 12th day, achieving a maximum bacterial population of $6.30x10^7$ Cell/mL in the test without the addition of $FeSO_4.7H_2O$, higher than the concentration of the initial inoculum compared to the concentration of the first stage. See **Figure 10**.

4.2.2 pH variation

The pH of the solution increases during the first 7 days, possibly due to the increase in pulp density, being controlled with sulfuric acid solution until it recovers

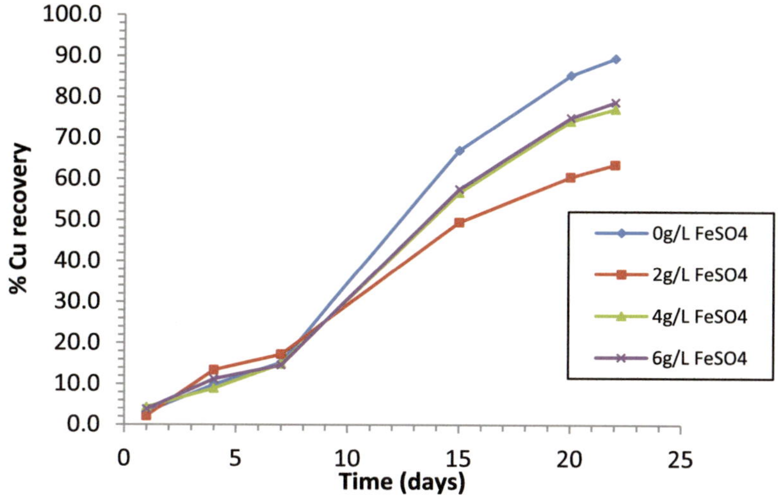

Figure 9.
Bioleaching in pulp containing 6% solids.

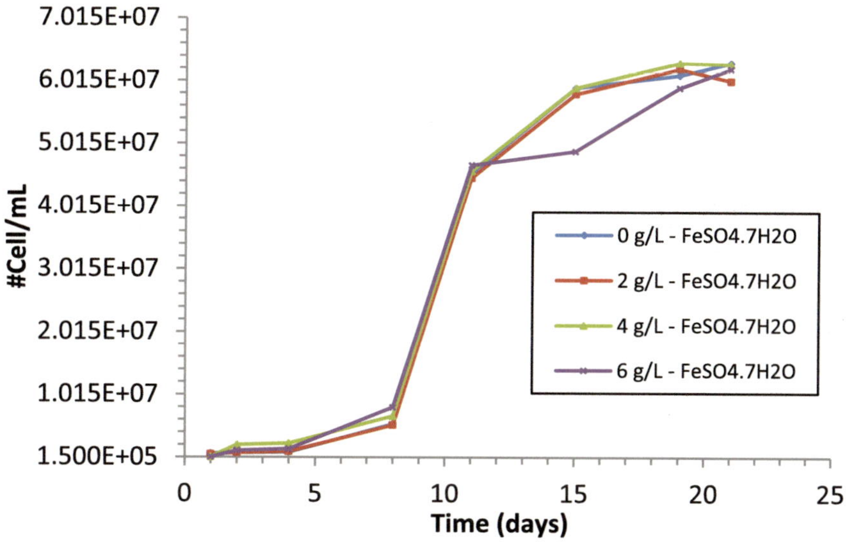

Figure 10.
Variation of the bacterial population during the second stage of sulfide ore bioleaching.

its initial value. As time goes by, the decrease is observed, being necessary for its recovery to the initial value of 1.8. See **Figure 11.**

4.2.3 Measurement of oxidation: Reduction potential

Figure 12 shows the behavior of the oxidation-reduction potential of all the tests. Achieving a maximum of 613.2 mV with the lowest amount of $FeSO_4.7H_2O$. It is also observed that they reach 600 mV approximately on the 8th day of leaching.

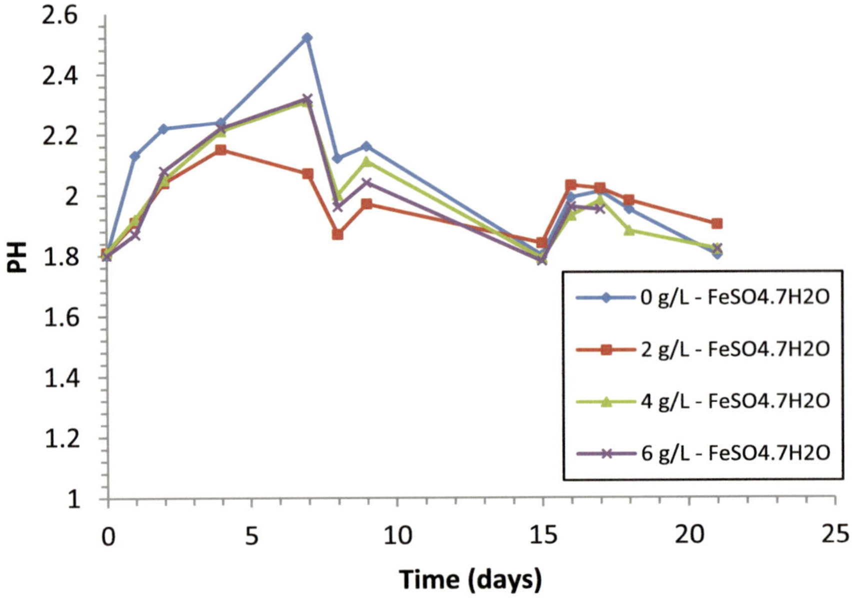

Figure 11.
pH variation during the second stage of bioleaching.

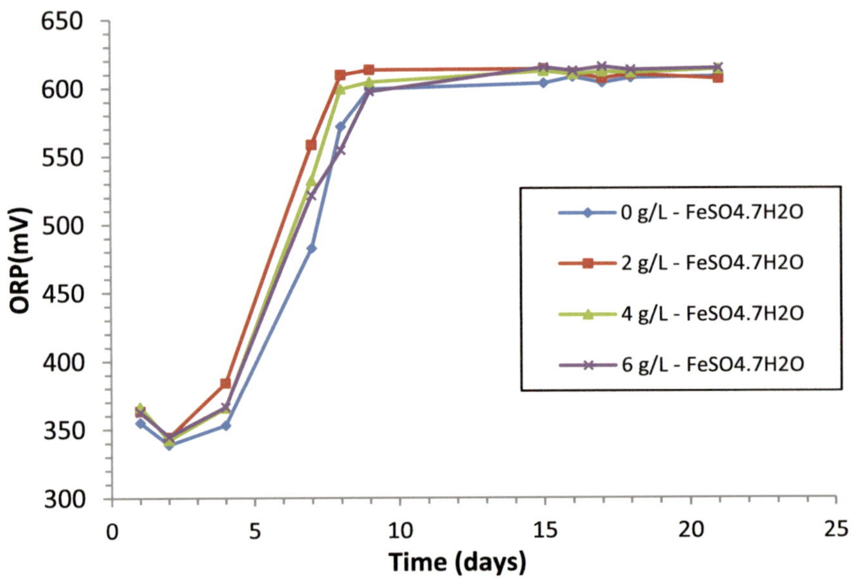

Figure 12.
Measurement of oxidation-reduction potential (ORP). Second stage of the bioleaching process.

4.3 Third stage of bioleaching

The use of Medium 9 k solution as leaching substrate is continued, varying the concentrations of $FeSO_4$ $7H_2O$. For the assays we continue using 500 mL Erlenmeyer flasks, add 6.0 g of sulfide mineral (2% W/V), 30 mL of bacterial strain (10% V/V), and 300 mL of 9 k substrate containing 0.0, 3.0, 9.0 and 15.0 g/L of $FeSO_4.7H_2O$.

The pH is regulated with sulfuric acid solution. The process was continued on a stirring platform at 150 rpm. The inoculum is obtained from the previous stage. Controls of pH, oxidation-reduction potential, and bacterial population are carried out.

After 25 days of leaching, an extraction of 85.6% copper was achieved in the test without the addition of ferrous sulfate. See **Figure 13**.

4.3.1 Effect on bacterial growth

In the tests, they were inoculated with the strain with the highest bacterial population resulting in the effluents of the second stage, whose bacterial population was $6.30x10^7$ Cell/mL. Exponential growth is achieved approximately after the second day of experimentation. In contrast to the first stage, the reduction was achieved in 8 days and compared to the second stage, in 5 days. Exponential growth ends after approximately 10 days. Higher growth ($4.51x10^7$ Cell/mL) is achieved in the test lacking the ferrous salt. It is concluded that bacterial adaptation and growth with the provision of mineral as a nutrient-supplying medium is a chemolithotrophic characteristic of the bacterium *Acidithiobacillus ferrooxidans*. See **Figure 14.**

4.3.2 pH variation

Increases and decreases in pH were observed during the first three days. The solution with 15 g/L of $FeSO_4.7H_2O$ reaches a pH of 2.08 and is corrected with sulfuric acid solution, then an increase in acidity is observed, reaching a pH of 1.3, possibly

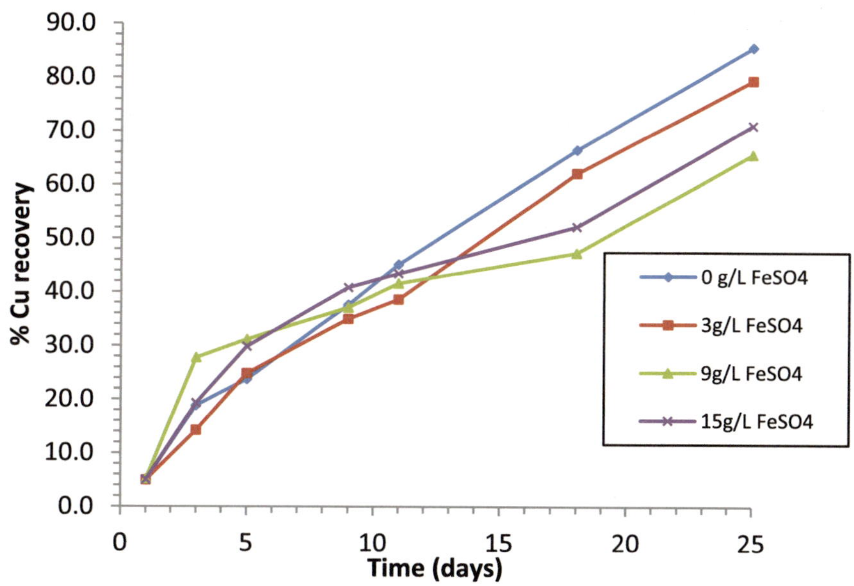

Figure 13.
Bioleaching in pulp containing 2% solids.

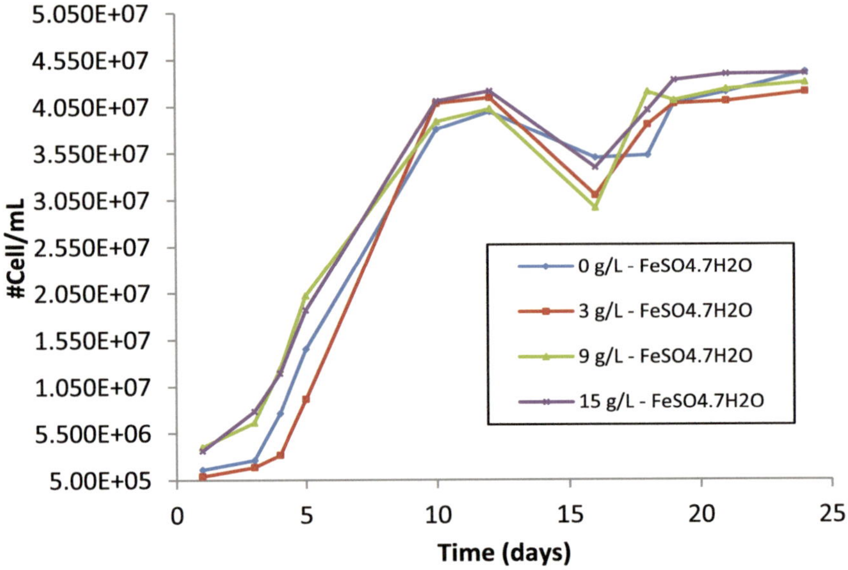

Figure 14.
Growth of the bacterial population. Third stage of experimental processing.

caused by the solubilization of the acid components of the mineral. In the following 17 days approximately, the variation is lower and is controlled with sulfuric acid solution, seeking to maintain around 1.8, then a marked decrease is observed (**Figure 15**).

4.3.3 Measurement of oxidation: Reduction potential

As can be seen in **Figure 16**, on the third day values close to the maximum are obtained, remaining almost constant during the rest of the test period. In contrast to the first stage where growth occurs between days 7 to 15 approximately, and in the second stage growth occurs between days 4 to 8 approximately. In this stage, the average maximum values oscillate around 585 mV for each of the tests.

5. Results

The level of adaptation to the new conditions is proportional to the amount of reseeding carried out and to the conditions of the substrates to which they are subjected with the possibility of making modifications. In the adaptation stage, the highest population growth of the bacteria isolated from the recovered mining unit is determined by the iron sulfate content in the substrate and the strict control of pH. These values were 22.2 g/L and 1.8, respectively.

Jarosite formation can occur at different potentials. The study by Ghahremaninezhad et al. [24], in several electrochemical circuits, identifies the formation of CuS at potentials around 750 mV and at 1400 mV the formation of jarosite, consequently, the hindering of the process. Yang et al. [23], in dissolution of a chalcopyrite electrode at potentials between 748 and 828 mV identified electrode passivation. In the present study, copper dissolution occurs throughout the test period and at the potentials revealed at each of the stages.

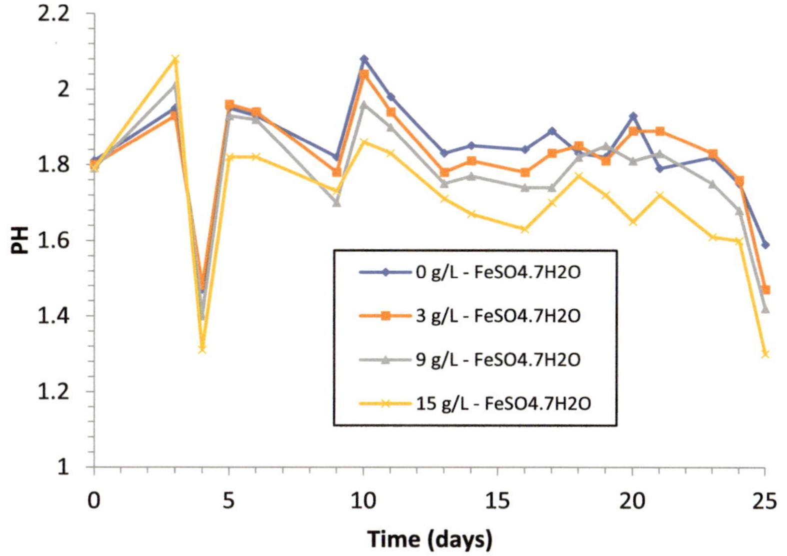

Figure 15.
pH variation during the third stage of bioleaching.

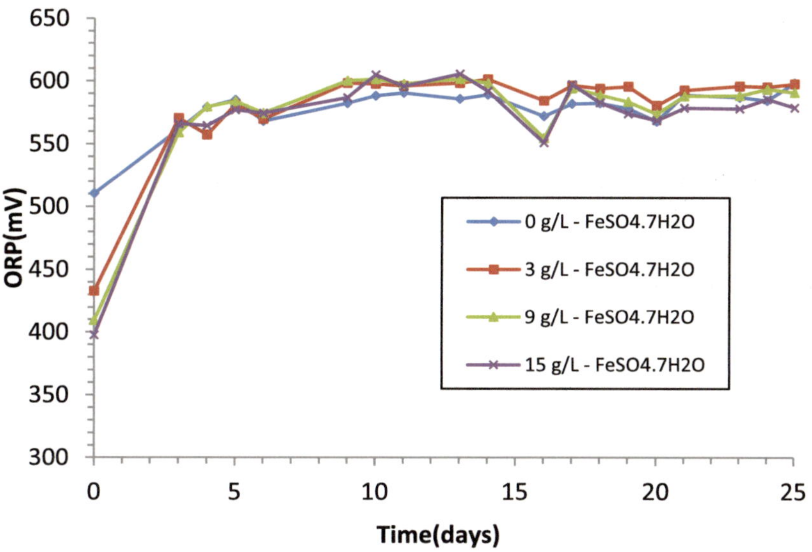

Figure 16.
Measurement of oxidation-reduction potential (ORP). Third stage of the bioleaching process.

In the first stage of bioleaching, population growth is achieved approximately in the period from the 12th to the 24th day, followed by a break and with a tendency to remain constant during the duration of the stage. The highest bacterial population was $4.75x10^7$ Cell/mL after 24 days with 6.0 g/L of $FeSO_4.7H_2O$ substrate and reaching only 67% of the initial inoculum. While the oxidation-reduction potential shows a varied behavior during the growth period. In the first 6 days, the sample without the ferrous salt increases from 360 mV to 585 mV on about the 10th day. The sample of 15 g/L of $FeSO_4.7H_2O$ has a delayed increase but reaches a maximum of 560 mV. The remaining samples, on average, reach 575 mV.

In the second stage, the inoculated strain had a concentration of 4.75x10^7 Cell/mL, the adaptation and growth phases were observed to have the same growth trend in all tests; the exponential phase began on the eighth day, reaching a maximum of 6.30x10^7 Cell/mL with 0.0 g/L of $FeSO_4.7H_2O$. The bacterial concentration is 42% higher in relation to the inoculum. The maximum value of oxidation-reduction potential is 613.2 mV. In the test, with 2 g/L of $FeSO_4$ $7H_2O$, it is observed that it exceeds 600 mV.

In the third stage, the beginning of exponential bacterial growth occurs on the third day after the start of bioleaching, 9 days shorter than in the first stage and 5 days shorter than in the second stage. The exponential growth ends after approximately 10 days. At this stage, it is observed that, in the absence of the ferrous salt, the concentration of 4.51x10^7 Cell/mL is achieved with a certain similarity to other concentrations of the iron salt. Meanwhile, ORP values have a remarkable evolution during the first 3 days and then tend to remain constant throughout the test period with an average value of 585 mV, 15 mV lower than in the second stage and 10 mV higher compared to the first stage.

Compared to the ORP increase, in the first stage, it occurs between days 6 to 10 approximately, in the second stage it occurs between days 4 to 8 and in the third stage it occurs in the first 3 days; with redox potential increases between 420 and 560 mV approximately. Contrasted with the results of the instrumental analysis carried out to determine the presence of chalcopyrite, disulfides, and polysulfides on the surface of the mineral causing the passivation and hindering the dissolution of the mineral, several redox potential ranges are identified. Thus, chalcopyrite is predominantly oxidized to polysulfide when the redox potential is below 350 mV and a low dissolution rate occurred when the redox potential is in the range of 350–480 mV, chalcopyrite was mainly transformed into Cu_2S intermediate species instead of polysulfide, increasing the dissolution rate, and when the redox potential is above 480 mV, chalcopyrite was directly oxidized to polysulfide, which causes passivation of chalcopyrite [3]. Also mentioned is the dissolution of iron from the chalcopyrite surface in the 475 to 700 mV potential range, leaving a slowly dissolving S_2^{2-} and S_n^{2-}, layer above 700 mV [38].

The measurement of the potential (Ev) is the dissolution of the electron giver and electron acceptor at varying substrate concentrations at pH 1.8 and at 22°C, showing increasingly positive values due to the increasing tendency to accept electrons with the consequent formation of sulfates. In this regard, Vilcáez et al. [21] mention that optimal temperatures for thermophile growth did not always mean high copper extraction yields, suggesting that with a high pH (pH 2.0), the bioleaching of chalcopyrite is more efficient, concluding that the bioleaching of chalcopyrite is controlled by ORP rather than by pH or temperature.

6. Conclusions and recommendations

The acid drainage of the mine workings studied (Huancavelica – Peru) is acidic, with a pH in the range of 3.0 to 4.5 pH, with a significant amount of metals in solution and abundant microorganisms such as the bacterium At. ferroooxidans.

Bacterial species are satisfactorily adapted to different media containing varying amounts of iron as sulfides and oxides, coming from highly mineralized quarries (presence of iron, copper, lead, zinc, sulfur, silica, gold, silver, and others). However, the qualitative and quantitative determination of the bacterial strain is still under investigation and will depend on the constitution of the mineral substrate provided.

The redox potential as a determinant of the growth and metabolism of the culture indicates its capacity to accept or donate electrons, that is, the oxidizing or reducing characteristics of the components of the medium or substrate, determined in part by the oxygen concentration. These oxidizing characteristics are those required by bacteria of the genus *thiobacillus*, favoring their growth and the development of an oxidative metabolism.

Also, the redox potential indicates the metabolic activities of living microorganisms and can be used to specify the environment in which microorganisms are able to generate energy and synthesize their enzymes or generate new cells without resorting to molecular oxygen.

Undoubtedly, the mineralogical composition of the mineral, as well as the structure of the species, together with the temperature, pH, and physical conditions of the mineral, will determine the bacterial growth, the redox potential, and the degree of dissolution and extraction of the elements of interest during the leaching process with sulfur and iron oxidizing microorganisms.

In the bioprocesses applied to sulfur minerals, the simple and compound ions, together with the bacterial consortium, transfer the electrons coming from the oxidation of inorganic matter to the available electron acceptors of a more oxidizing nature, allowing to obtain the greatest margin of energy gain for the oxidation of the mineral substrate present, from which the carbon and energy necessary for its evolution are provided, being a mechanism typical of chemo lithotrophic organisms.

The oxide reduction potential offers many advantages in real-time monitoring. The variation in the dissolution of mineral sulfides can be attributed mainly to two factors, 1. the type of measuring electrode and 2. the composition of the mineral substrate, the dissolution of some of its components will determine the change in pH and consequently the increase or decrease of the potential.

The different oxidation statuses of sulfur (−2, 2, 2, 4, and 6) provide a redox potential and a great variety of enzymes that can oxidize different inorganic sulfur compounds; for this reason, it is advisable to identify them, as well as the metabolic routes, allowing to optimize the conditions of the sulfur oxidation reactions and to improve the bacterial catalytic activity.

During bioleaching, after a period of time, the oxidation rate and/or the dissolution of the valuable elements present in the ore show a decrease or even interruption caused by the passivation of the ore surface, as well as by the saturation of the medium with ionic compounds. Therefore, it is recommended to purify or change the enriched solution.

Acknowledgements

The authors would like to thank the Vice-Rectorate for Research of the Universidad Nacional Mayor de San Marcos (UNMSM); which, through the Superior Research Council, financed the execution of projects C17162131, C18160202, C19161651 and C23160791, of the Research Group Clean Technologies for Environmental Coexistence (TELICMA); likewise, the students and teachers of the Faculties of Biological and Chemical Sciences and Chemical Engineering, who participate in the achievement of the objectives of the research group.

Author details

Vladimir Arias-Arce[1*], Daniel Lovera-Dávila[1], José J. Guerrero-Rojas[2], Fanny Blas-Rodriguez[3] and Ismael Molina-Pereyra[4]

1 Metallurgical Engineering Deparment, FIGMMG-UNMSM, Lima, Perú

2 Universidad Privada Norbert Wiener, Lima, Perú

3 Chemical Engineering Deparment, FQ&IQ - UNMSM, Lima, Perú

4 Unidad de Posgrado, FIGMMG-UNMSM, Lima, Perú

*Address all correspondence to: variasa@unmsm.edu.pe

References

[1] Majima H. How oxidation affects selective flotation of complex sulphide ores. Canadian Metallurgical Quarterly. 1969;**8**(3):269-273. DOI: 10.1179/cmq.1969.8.3.269

[2] Fagundo Castillo JR, González Hernández P, Suárez Muñoz M, Melián Rodríguez C. Relaciones entre potenciales redox y concentraciones de sulfuros en aguas termales de Cuba. Contribución a la educación y protección ambiental. 2005;**6**:31-44. Available from: https://www.researchgate.net/profile/Patricia-Gonzalez-Hernandez/publication/

[3] Wang J, Gan X, Zhao H, Hu M, Li K, Qin W, et al. Dissolution and passivation mechanisms of chalcopyrite during bioleaching: DFT calculation. XPS and electrochemistry analysis. Minerals Engineering. 2016;**98**:264-278. DOI: 10.1016/j.mineng.2016.09.008

[4] Alvarez M. Microbial Treatment of Heavy Metal Leachates. Spain: Gráficas Terrasa. Department of Biotechnology, Lund University; 2009. Available from: https://lup.lub.lu.se/search/publication/24654

[5] Arias-Arce VA, Lovera Dávila DF, Paucarima AF, Rojas TL. Correlación del potencial óxido reducción y la población bacteriana durante el estudio de biolixiviación de sulfuros de cobre. Revista del Instituto de investigación de la Facultad de minas, metalurgia y ciencias geográficas. 2021;**24**(47):19-28. DOI: 10.15381/iigeo.v24i47.20639

[6] Kaksonen AH, Plumb JJ, Franzmann PD, Puhakka JA. Simple organic electron donors support diverse sulfate-reducing communities in fluidized-bed reactors treating acidic metal- and sulfate-containing wastewater. FEMS Microbiology Ecology. 2004;**47**(3):279-289. DOI: 10.1016/S0168-6496(03)00284-8

[7] Zhang L, Wu J, Wang Y, Wan L, Mao F, Zhang W, et al. Influence of bioaugmentation with Ferroplasma thermophilum on chalcopyrite bioleaching and microbial community structure. Hydrometallurgy. 2014;**146**:15-23. DOI: 10.1016/j.hydromet.2014.02.013

[8] Yang C, Jiao F, Qin W. Co-bioleaching of chalcopyrite and silver-bearing Bornite in a mixed moderately thermophilic culture. Minerals. 2017;**8**(1):4. DOI: 10.3390/min8010004

[9] Li J, Yang H, Tong L, Sand W. Some aspects of industrial heap bioleaching technology: From basics to practice. Mineral Processing and Extractive Metallurgy Review. 2022;**43**(4):510-528. DOI: 10.1080/08827508.2021.1893720

[10] Sand W, Gerke T, Hallmann R, Schippers A. Sulfur chemistry, biofilm, and the (in)direct attack mechanism a critical evaluation of bacterial leaching. Applied Microbiology and Biotechnology. 1995;**43**(6):961-966. DOI: 10.1007/BF00166909

[11] Rohwerder T, Sand W. Mechanisms and biochemical fundamentals of bacterial metal sulfide oxidation. Microbial Processing of Metal Sulfides. 2007:35-58. DOI: 10.1007/1-4020-5589-7

[12] Anjum NA, Ahmad I, Mohmood I, Pacheco M, Duarte AC, Pereira E, et al. Modulation of glutathione and its related enzymes in plants' responses to toxic metals and metalloids—A review. Environmental and Experimental

Botany. 2012;**75**:307-324. DOI: 10.1016/j.envexpbot.2011.07.002

[13] Fu KB, Lin H, Mo XL, Wang H, Wen HW, Wen ZL. Comparative study on the passivation layers of copper sulphide minerals during bioleaching. International Journal of Minerals, Metallurgy, and Materials. 2012;**19**:886-892. DOI: 10.1007/s12613-012-0643-x

[14] Crundwell FK. The semiconductor mechanism of dissolution and the pseudo-passivation of chalcopyrite. Canadian Metallurgical Quarterly. 2015;**54**(3):279-288. DOI: 10.1179/1879139515Y.0000000007

[15] Zhao H, Zhang Y, Zhang X, Qian L, Sun M, Yang Y, et al. The dissolution and passivation mechanism of chalcopyrite in bioleaching: An overview. Minerals Engineering. 2019;**136**:140-154. DOI: 10.1016/j.mineng.2019.03.014

[16] Espinoza J, Revah S, Le Borgne S. Rutas metabólicas de oxidación del azufre en bacterias quimiolitoautótrofas, relevancia ambiental y biotecnología. Mensaje Bioquímico. 2010;**XXXIV**:101-120. Available from: http://bq.unam.mx/mensajebioquimico

[17] Wang J, Zhu S, Zhang YS, Zhao HB, Hu MH, Yang CR, et al. Bioleaching of low-grade copper sulfide ores by Acidithiobacillus ferrooxidans and Acidithiobacillus thiooxidans. Journal of Central South University. 2014;**21**(2):728-734. DOI: 10.1007/s11771-014-1995-3

[18] Hiroyoshi N, Miki H, Hirajima T, Tsunekawa M. A model for ferrous-promoted chalcopyrite leaching. Hydrometallurgy. 2000;**57**(1):31-38. DOI: 10.1016/S0304-386X(00)00089-X

[19] Nazari B, Jorjani E, Hani H, Manafi Z, Riahi A. Formation of jarosite and its effect on important ions for Acidithiobacillus ferrooxidans bacteria. Transactions of Nonferrous Metals Society of China. 2014;**24**(4):1152-1160. DOI: 10.1016/S1003-6326(14)63174-5

[20] Zhao H, Wang J, Yang C, Hu M, Gan X, Tao L, et al. Effect of redox potential on bioleaching of chalcopyrite by moderately thermophilic bacteria: An emphasis on solution compositions. Hydrometallurgy. 2015;**151**:141-150. DOI: 10.1016/j.hydromet.2014.11.009

[21] Vilcáez J, Suto K, Inoue C. Bioleaching of chalcopyrite with thermophiles: Temperature–pH–ORP dependence. International Journal of Mineral Processing. 2008;**88**(1-2):37-44. DOI: 10.1016/j.minpro.2008.06.002

[22] Liu HC, Xia JL, Nie ZY. Relatedness of Cu and Fe speciation to chalcopyrite bioleaching by Acidithiobacillus ferrooxidans. Hydrometallurgy. 2015;**156**:40-46. DOI: 10.1016/j.hydromet.2015.05.013

[23] Yang Y, Harmer S, Chen M. Synchrotron-based XPS and NEXAFS study of surface chemical species during electrochemical oxidation of chalcopyrite. Hydrometallurgy. 2015;**156**:89-98. DOI: 10.1016/j.hydromet.2015.05.011

[24] Ghahremaninezhad A, Asselin E, Dixon DG. Electrochemical evaluation of the surface of chalcopyrite during dissolution in sulfuric acid solution. Electrochimica Acta. 2010;**55**(18):5041-5056. DOI: 10.1016/j.electacta.2010.03.052

[25] Zhao H, Wang J, Tao L, Cao P, Yang C, Qin W, et al. Roles of oxidants and reductants in bioleaching system of chalcopyrite at normal atmospheric pressure and 45 C. International Journal

of Mineral Processing. 2017;**162**:81-91. DOI: 10.1016/j.minpro.2017.04.002

[26] Khoshkhoo M, Dopson M, Shchukarev A, Sandström Å. Chalcopyrite leaching and bioleaching: An X-ray photoelectron spectroscopic (XPS) investigation on the nature of hindered dissolution. Hydrometallurgy. 2014;**149**:220-227. DOI: 10.1016/j.hydromet.2014.08.012

[27] Johnson DB. The biogeochemistry of biomining. In: Barton L, Mandl M, Loy A, editors. Geomicrobiology: Molecular and Environmental Perspective. 2010. pp. 401-426. DOI: 10.1007/978-90-481-9204-5_19

[28] Arias V, Lovera D, Quiñones J, Flores A, Gil J, Ramírez L, et al. Biolixiviación de cobre a partir de minerales sulfurados con altos tenores de pirita y arsenopirita. Revista Del Instituto de Investigación de La Facultad de Ingeniería Geológica, Minera, Metalurgica y Geográfica. 2015;**18**(36):157-164. DOI: 10.15381/iigeo.v18i36.12164

[29] Acevedo F, Gentina J. Biolixiviacion de minerales de cobre. Fundamentos y Perspectivas de las Tecnologías Biomineras. Archivos de Ingeniería Bioquímica. 2005:45-61. Available from: https://docplayer.es/10211791-Fundamentos-y-perspectivas-de-las-tecnologias-biomineras.html

[30] Hallberg K, Johnson B. Novel Acidophiles isolated from moderately acidic mine drainage waters. Hidrometallurgy. 2003;**71**(1):139-148. DOI: 10.1016/S0304-386X(03)00150-6

[31] Arias V, Rodríguez C, Ramírez P, Nonones E, Salazar D, Gil J, et al. Aislamiento de bacterias acidófilas a partir del drenaje ácido proveniente de las inmediaciones a las unidades mineras de Julcani y Recuperada, Huancavelica. Revista del Instituto de investigación de la Facultad de minas, metalurgia y ciencias geográficas. 2012;**15**(30):59-66. DOI: 10.15381/iigeo.v15i30.3450

[32] Akcil A, Ciftci H, Deveci H. Role and contribution of pure and mixed cultures of mesophiles in bioleaching of a pyritic chalcopyrite concentrate. Minerals Engineering. 2007;**20**:310-318. 2. DOI: 10.1016/j.mineng.2006.10.016

[33] Ospina JD, Mejía Restrepo E, Osorno Bedoya L, Márquez MA, Morales AL. Biooxidación de concentrados de arsenopirita por Acidithiobacillus ferrooxidans en erlenmeyer agitados. Revista Colombiana de Biotecnología. 2012;**XIV**(1):135-145. Available from: https://revistas.unal.edu.co/index.php/biotecnologia/article/view/31851#textoCompletoHTML

[34] Arias V, Anaya F, Quiñones J, Salazar, D, Gil J, Jamanca G. Adaptación del Thiobacillus Ferrooxidans a sustratos conformados con especies de minerales piríticos. Revista del Instituto de investigación de la Facultad de minas, metalurgia y ciencias geográficas. 2013;**16**(31):40-46. DOI: 10.15381/iigeo.v16i31.8339

[35] Rivera R, Camejo P, Moya F, López J, Munguía M. Estudio de biolixiviación de un mineral de sulfuros de un mineral de sulfuros de cobre de baja ley con bacterias Tio- y Ferro-oxidantes en condiciones termófilas. Revista de La Facultad de Ingeniería. 2011;**26**:65-73. Available from: http://www.revistaingenieria.uda.cl/Publicaciones/260009.pdf

[36] Pradhan N, Nathsarma KC, Srinivasa Rao K, Sukla LB, Mishra BK. Heap bioleaching of chalcopyrite: A review. In Minerals Engineering. 2008;**21**(5):355-365. DOI: 10.1016/j.mineng.2007.10.018

[37] Rawlings DE. Characteristics and adaptability of iron- and sulfur-oxidizing microorganisms used for the recovery of metals from minerals and their concentrates. In Microbial Cell Factories. 2005;**4**:1-15. DOI: 10.1186/1475-2859-4-13

[38] Yang C, Qin W, Zhao H, Wang J, Wang X. Mixed potential plays a key role in leaching of chalcopyrite: Experimental and theoretical analysis. Industrial & Engineering Chemistry Research. 2018;**57**(5):1733-1744. DOI: 10.1021/acs.iecr.7b02051

Chapter 7

Role of Various Physicochemical Factors in Enhancing Microbial Potential for Bioremediation of Synthetic Dyes

Radhika Birmole and Aruna K. Samudravijay

Abstract

The Indian dye industry is globally recognized for production and export of every known class of dye. On the less attractive side of industrialization, they contribute considerably to environmental pollution. The dyes discarded by industries persist in the environment due to extremely slow rate of biodegradation. Moreover, these dyes are toxic to insects, birds and terrestrial life. The dyes also hamper the light penetration in water bodies, severely affecting the the process of photosynthesis. In spite of the problems associated with synthetic dye disposal, they are industrially preferred due to their fundamental requirement in enhancing overall appearance of goods, quality and cost effectiveness. Several studies have reported physicochemical techniques for remediation of dye effluents. Most of these techniques pose significant drawbacks due to their high energy and cost requirements. The bioremediation approach, on the other hand, offers advantages of sustainable environmental friendly processes to detoxify and degrade dyes into harmless products. This chapter provides an overview of the potential role of various physicochemical factors such as pH, temperature, oxygen and nutrient concentration in optimum decolorization of dyes by naturally isolated microbial strains. In addition, the role of cosubstrates, electron acceptors and microbial enzymes are also discussed.

Keywords: sustainable, bioremediation, physicochemical techniques, dyes, optimum

1. Introduction

The synthetic textile industry is estimated to be a trillion dollar industry that utilizes dyes on a large scale. Besides, dyes are used extensively in most of the small and large scale paint, automobile, and cosmetic industries [1]. According to the most recent buiseness report, several thousand tons of dyes are manufactured every year acquiring 15.59 billion dollar global market in 2022 with a compound annual growth rate of 11.5% [2]. Statistically, among the manufactured dyes, 70% are azo-dyes and 15% are anthraquinone dyes. The remaining 15% constitute of all other dyes such as triphenylmethane and phthalocyanine dyes [3, 4]. It is estimated that 20% of the dyes are lost during processing and application and become a part of industrial effluents [1]. The synthetic dyes are derived from petroleum products. They contain unsaturated chromophore groups that contribute to the color and stability of the molecule. Many of the synthetic dyes are reported to have carcinogenic properties. Disposal of untreated dye effluents significantly harms the soil as well as aquatic flora [5]. For this reason, remediation processes have received tremendous attention from researchers of related fields.

The physical techniques of dye remediation include reverse osmosis, photodegradation, coagulation, flocculation, and ion exchange method for ultra-filtration. However, these processes generate huge amount of sludge, which contains a mixture of partially remediated and precipitated compounds. They are collectively known as secondary waste products [6]. Most of these secondary waste compounds produced on photodegradation exhibit equal or more toxicity as compared to the parent dye molecule [7, 8]. As opposed to physical methods, the biological techniques for the degradation of dyes have three major advantages [9, 10].

1. Bioprocess technologies do not contribute to secondary contamination. There is relatively less sludge production and degraded dye metabolites are environmentally benevolent.

2. Biological methods can be implemented in situ at the polluted site.

3. They are cost effective.

The environment is replete with a wide range of microorganisms possessing unique abilities to degrade natural and man-made complex compounds. The screening and isolation of microorganisms degrading synthetic dyes from a variety of natural samples or microorganism enriched samples due to anthropogenic activities have been carried out by many researchers. Bacteria are the most frequently preferred microorganisms for treating industrial effluents because they are easily adapted to new environment and have short generation time. Some being facultative can grow aerobically or anaerobically. Some bacteria have naturally adapted to live in severe environmental settings with respect to salinity, pH, and temperature and have the ability to produce a variety of oxidoreductases that effectively remediates several classes and mixture of dyes [9].

The general observation by many workers is that most azo dyes resist biodegradation in traditional aerobic sewage treatment plants due to inhibition of azoreductase [11–16]. From the existing available knowledge, bacterial anaerobic azo bond cleavage is well suited in the elimination of azo dyes from effluent processing systems. In static cultures, the exhaustion of oxygen can be easily achieved for the purpose of breaking the azo linkage by facultative and anaerobic bacteria. These biochemical reactions occur at pH 7.0 and involve low-molecular-weight redox mediators, which are extremely unspecific. The existence of supplementary carbon nutrients elevates the reduction rates. The reducing counterparts are created during metabolism of these nutrients, which can ultimately be utilized for reducing the azo dye biochemically [14, 15]. The production of naphthylamine, benzidine derivatives, and other amines, however, is a major limitation in anaerobic biodegradation of azo dyes due to their presumed toxicity and serious health threat to humans [11, 13]. As a practical solution to this problem, comparatively few reports of azoreductase production by aerobic bacteria are reported in literature [17, 18]. For instance, Oturkar *et al.* [19] investigated the decolorization of RR120 using *Bacillus lentus* and reported 98% and 70% decolorization under anoxic and aerobic conditions respectively. However, the presence of N2 gas with complete O2 free atmosphere led to loss of decolorization ability of *B. lentus*. Hence, screening of potential bacterial strains and optimization of their dye degradation ability can be extremely helpful in overcoming the challenges of dye remediation.

2. Selection of promising isolates

The original load of microorganisms capable of degrading dye molecules is low in natural samples. Hence, many research groups carry out enrichment of samples during screening programs using growth media such as mineral salts medium or Bushnell and Hass medium (BHM), either amended or not amended, with organic carbon and/or nitrogen sources [20–28], medium containing peptone-yeast extract or peptone-glucose [27, 29], nutrient broth [30–32], and synthetic wastewater [33]. A different formulation of medium consisting of 1% peptone in distilled water/seawater in 1:1 combination was used by Srinivasan *et al.* [34] for the enrichment of mangrove sediments.

Also, considering the diversity of microbial species in natural ecosystem, the possibility of isolation of potential dye degraders is increased on careful selection of samples. Few examples of samples used by scientists for isolation of dye degraders are mangrove swamps [35], water and sludge from the drain [20], soil collected from distillery spent wash contaminated sites [36], textile dye effluent from CETP (Common Effluent Treatment Plant), Perundurai, Chennai, Tamilnadu [37], soil near the tannery, Central Leather Research Institute, Tamil Nadu, Chennai, India [38], the effluent from dye manufacturing unit [31], lake mud [39], soil sample from tannery effluent [32], soil, textile effluent and sewage [21], and beach water [40].

For selection and purification of the promising isolates for dye decolorization/degradation process, enriched microflora is isolated on the solid media of same or different composition with or without synthetic dye. Sugiura *et al.* [41] made use of synthetic medium comprising azo dyes as the sole carbon supply for enrichment of soil samples and isolated dye decolorizing bacteria from enriched media on nutrient agar containing 0.02% azo dye. BHM supplemented with dye (Direct Red 81, 100 ppm) being a single carbon source or accompanied by yeast extract and glucose was used for the enrichment and isolation of samples obtained from effluent-contaminated areas near-by dyestuff units [24]. In contrast, Alhassani *et al.* [42] used nutrient broth and nutrient agar without dye for the enrichment and isolation during screening.

During the screening process, microorganisms are assessed based on their ability to decolorize the dye metabolites. Confirmation of dye degradation is done by evaluating the nature of degraded dye metabolites using analytical techniques such as High-Performance Liquid Chromatography (HPLC) or Gas Chromatography Mass Spectrophotometry (GC-MS). The two different approaches for selection of efficient dye decolorizer degrader are reported in literature. First approach uses nutrient medium without dye for isolation, and morphologically distinct colonies are selected for further dye decolorization assay in liquid culture medium having defined dye concentration. The quantitative measurement of the decolorization given by the organisms in terms of either percentage decolorization or the rate of decolorization is considered for the selection of strains [30]. The preliminary quantitative measurement of dye decolorization/degradation is estimated using a UV-Visible spectrophotometer. The decolorized broth is freed of cells and used for spectrum scan from 200 to 800 nm using sterile nutrient medium as blank. Abiotic control is also subjected to spectrum scan using same range of wavelengths. The absorbance values of abiotic control at the λ_{max} of the dye in the visible range and that of the decolorized broth at the same λ_{max} are used to calculate the percent decolorization [43]. Based on the above method, Guadie *et al.* [22] isolated and purified 135 morphologically different colonies using spread plate method and tested decolorization/degradation in liquid mineral salts medium containing RR239. Similarly, Kannan *et al.* [24] obtained seven isolates on nutrient agar plates, which were further grown in mineral salts medium containing Remazol Black B dye (1000 mg/L) for assessing the decolorization/degradation ability.

The second approach used for selection of decolorizers uses isolation of the enriched samples on dye containing nutrient agar, to observe clear zones around the colonies. The organisms showing the largest decolorization zones are selected for further screening of their decolorization potential. Using this approach, Chen *et al.* [39] observed white and pink colonies with decolorized zones on screening medium containing RED RBN dye, and these were further tested for decolorization capability in submerged cultures.

In contrast to the above two approaches, Nawahwi *et al.* [44] described a rapid isolation technique without enrichment to isolate the dye decolorizing bacteria on sterilized textile wastewater agar medium (wastewater sterilized using membrane filter or autoclaved) using spread plate technique. The bacteria that formed colonies with clear zones on this medium were selected for further screening. This method was also applied for the decolorization by *Basidiomycetes* strains on agar plates containing 0.2 g/L of Orange G or Remazol Brilliant Blue R; however, only 15 strains decolorized both the dyes tested with demarcated clear regions surrounding the colonies [45].

3. Inoculum preparation

The commonly used biomass parameters, i.e., Optical Density (OD) and volume of cells (saline suspension or medium grown) are used to standardize the optimum cell number for a successful decolorization/degradation process. Various researchers describe a number of ways for the preparation of inoculum. Few studies have employed overnight grown pure cultures for studying dye decolorization as described by Mahmood *et al.* [20] wherein 0.6 OD_{597nm} was adjusted for five bacterial isolates to optimally decolorize Remazol Black B. For the biodecolorization of reactive orange using *Bacillus* sp. ADR, 1.0 OD_{620nm} equivalent to 1.818 g dry cell weight/L was used by Telke *et al.* [46]. A 10% v/v of freshly grown *Brevibacillus* sp. with 0.3 OD_{600nm} was used in degradation of Toluidine Blue [41].

Khalid *et al.* [25] prepared inocula of two *Shewanella* sp. with 1.0 OD_{550nm} and used 2%v/v for azo dye decolorization. In another study, Junnarkar *et al.* [23] used inoculum size 5–30% v/v with 5% interval of novel bacterial consortium for Direct Red 81 degradation. Similar inoculum preparation was reported by Du *et al.* [47] for *Aeromonas* sp. strain DH-6 during metabolism of azo dye Methyl Orange. Likewise, the ratio of 50:1 of medium to inoculums (OD_{600nm} 1.0) was used for studying degradation of textile dyes by *Stenotrophomonas maltophila* RSV-2 [48].

Similarly, the inoculum size of 4–20% at intervals of 4%v/v was used for the bacterial consortium RVM11.1 in the decolorization of RV 5 [26]. Kurade *et al.* [49] have reviewed biodegradation of Disperse Red 54 with pregrown cell mass (24 h) of *Brevibacillus laterosporus*. Inoculum size of 1–5% v/v was tested using 3-day-old broth grown *Streptomyces* DJP15 for the degradation of Azo Blue dye [50]. Similarly, *Shewanella decolorationis* S12 was grown in 50 mL LB on shaker (150 rpm) at 30°C. The cell mass was collected by centrifugation and uniformly suspended in phosphate buffer, pH 8.0, which was used as 10%v/v inoculum size of suspension for all the biodegradation experiments [51].

4. Physicochemical parameters influencing the decolorization/ degradation of dyes by bacterial isolates

4.1 Nutrient media

The dye degradation studies have used complex, synthetic, or semisynthetic nutrient media for studying biodegradation of dyes. The qualitative and quantitative nutrient composition of the medium is important for microbial degradation. This is due to the fact that certain nutrients of the medium are better donors of electrons than others to reduce the azo linkages in the substrate dye under static condition, or the nutrients may provide certain vitamins essential for catalytic activity of enzymes important during dye degradation. Khan [40] studied degradation of dye Red 2G using *Bacillus megaterium* sourced from seawater wherein a modified mineral salt basal medium deficient of nitrogen source was used to check the proficiency of the decolorizing strain to use the dye as a nitrogen source. However, most microorganisms require an organic cosubstrate for the dye degradation, thereby essentially necessitating the use of complex media for the degradation of the dye. For instance, Junnarkar *et al.* [23] used bacterial consortium NBNJ6, which exhibited best decolorization of Direct Red 81 dye when casein and starch were present in the medium. The bacterial cultures are incapable of decolorizing the dyes without cosubstrate/ electron donor, which suggests that the availability of additional carbon resource is vital for the growth of the bacteria as well as for the dye decolorization [27]. However, few microorganisms use dyes as the only source of nitrogen and carbon. One such evidence is reported for *Bacillus* sp. isolated in the degradation process of Congo Red by Gopinath *et al.* [32].

Jadhav *et al.* [52] have stated five media to understand the influence of different nutrients on methyl red degradation by *Saccharomyces cerevisiae* MTCC 463. These media were plain distilled water (100 mL), 5% mineral salts medium supplemented with 1% glucose, yeast and peptone extract at 0.1% each in distilled water, 0.1% peptone in distilled water, and 0.1% yeast extract in distilled water. The significance of certain nutrients for dye degradation can be interpreted with such experiments. The decolorization/degradation of sulfonated azo dye, Reactive Orange 16, in nutrient broth by *Bacillus* species ADR was studied by Telke *et al.* [46]. In the same year, Ben Mansour *et al.* [53] used mineral growth medium having a wide range of inorganic salts as macro and micronutrients and presence of glucose (1%) to study degradation of Acid Violet 7 by *Pseudomonas putida* mt-2. *E. coli* and *Pseudomonas* sp. were used for decolorization/degradation of Direct Black 38 and Congo Red (Direct Red 28) in Vanderbilt mineral medium containing glucose as cosubstrate. Similarly, the biodegradation of textile azo dyes such as Direct Blue 71, RB 5, RY107, and RR 198 by *Staphylococcus arelettae* strain VN-11 was studied in rich mineral salts medium containing a variety of salts, 0.3% glucose, 100 ppm of each dye with 0.1% yeast extract [54]. A model dye Direct Black 22 and few other azo dyes were studied for degradation by novel bacterial consortium DMC using Bushnell Hass medium supplemented with 0.1% glucose, 0.06% yeast extract (BGY medium) [14]. Usha *et al.* [21] studied degradation of RR120 and RB5 by *Aeromonas punctata* and *P. aeruginosa* using mineral salts medium broth and agar supplemented with 50 ppm glucose. Pandey and Dubey [37] used LB medium to degrade dye RR-BL using *Alcaligenes* sp. AA09. Also, the decolorization studies of Azo Blue dye were carried out in starch-casein broth with *Streptomyces* DJP15 by Pillai [50], while Modi *et al.* [31] carried out the decolorization assays for RR195 with pure culture and consortium using nutrient broth with and without glucose and in mineral salt medium.

4.2 Role of aeration

The dye degradation under static or shaker condition is closely based on the dye chemistry. The chemical structure and functional groups of the dye alter the efficiency of bacterial dye decomposition. Moreover, the decolorizing ability of the microorganism is closely related to its function as a facultative anaerobe. To understand the effect of aeration on dye degradation, facultative cultures such as *E. coli* and *Pseudomonas* species were grown aerobically, anaerobically, and microaerophilically using Direct Black 38 and Congo Red. Sodium thioglycollate was used to create anaerobiosis. Nachiyar and Rajakumar [38] subjected degradation assay of Navitan Fast Blue S5R by *P. aeruginosa* under shaker and static culture conditions. Pandey and Dubey [37] also carried out biodegradation study of RR-BL by *Alcaligenes* sp. AA09 under static condition and on an orbital shaker (150 rpm). Kalme *et al.* [55] too used static and shaking condition (120 rpm) with *Pseudomonas desmolyticum* NCIM 2112 to study biodegradation of Direct Blue 6. However, Modi *et al.* [31] subjected decolorization/degradation by bacterial isolates of four dyes soluble in water under static condition only. Elisangela *et al.* [56] evaluated decolorization of four azo dyes using *Brevibacterium* species strain VN-15 in an agitated and static sequential batch procedure. The choice for degradation of various dyes by microorganisms is not defined and changes under different conditions. Therefore, both anaerobic and aerobic conditions were used in a fluidized bed inoculated with *Pseudomonas* sp. L1 for biodegradation of Reactive Blue 13 experimented out by Lin *et al.* [57]. Similarly, disperse dye, Terasil Black degradation by recently isolated *Bacillus* species was tested on shaker (200 rpm) and under anaerobic condition. The anaerobic condition was created by sealing the flasks with sterile rubber stopper and purged with oxygen free nitrogen and incubated under stationary condition [58].

4.3 Role of temperature

The temperature affects dye degradation process significantly. Incubation temperature for decolorization/degradation ranging from 5 to 45°C was evaluated for degradation of Malachite Green by *Sphingomonas paucimobilis*. Telke *et al.* [46] also examined nearly same temperature range of 4–50°C while studying color removal of Reactive Orange 16 by *Bacillus* sp. ADR. A narrow range of temperatures between 25 and 50°C was used by Junnarkar *et al.* [23] while studying the degradation of Direct Red 81 with the help of a fresh bacterial consortium. Likewise, *Bacillus* sp. was subjected to incubation temperatures of 31–41°C at intervals of 2°C while degrading Congo Red [32]. Saratale *et al.* [59] studied the decolorization/degradation under the temperature range of 30–50°C for Reactive Green 19A degradation in nutrient broth with *Micrococcus glutamicus* NCIM-2168. Decolorization of Brilliant Blue G by fungal-bacterial consortium was carried out at 5°C, 30°C, 50°C, and 60°C [60]. Parshetti *et al.* [61] studied decolorization performance for *Kocuria rosea* MTCC 1532 with Methyl Orange using temperature range 10–50°C at intermissions of 10°C. *Aeromonas hydrophila* strain DN322 was inoculated in M-9 medium with 0.1% (w/v) yeast extract, which was incubated at 4°C, 10°C, 20–50°C at intervals of 5°C to establish the optimal temperature for its growth and decolorization of dyes [62]. Lastly, separate batches of Nutrient medium were incubated at 15–40°C for the biodegradation study of Crystal Violet by *Shewanella* species [63].

4.4 Role of pH

The acidity or alkalinity of the growing environment can alter the microbial multiplication rate and biochemical activities, thereby influencing the decolorization efficiency. Hence, an ideal pH is essential for the multiplication of the decolorizing strain as well as enhancing the decolorizing activity. This is because the dye degrading process is a metabolic process governed by enzymes. Thus, for the highest rate of decolorization, the optimal pH for the organism needs to be identified [64]. A range of pH from 2.0–14.0 was used for studying decolorization/degradation of Direct Red 5B in nutrient broth by *Comamonas* sp. UVS [65]. Similar range of pH in basal medium with molasses as carbon source was evaluated by Tamboli *et al.* [66] while studying the biodegradation of Direct Red 5B using *Sphingobacterium* sp. ATM. The biodisintegration of Scarlet R with the help of microbial consortium GR was checked in the pH range of 5.0–12.0 in nutrient broth by Saratale *et al.* [67]. Similarly, for the consortium of *Bacillus* species and *Galactomyces geotrichum*, pH 3.0, 5.0, 7.0, 9.0, and 12.0 were studied for Brilliant Blue G degradation [65]. On the same note, Li and Guthrie [68] studied metal-complex dyes removal by *Shewanella* strain J18143 at pH 4.0, 5.6, 6.8, 8.0, and 9.2. Khalid *et al.* [69] also used similar pH range of 4.0–10.0 in mineral salt medium for the azo dyes degradation by employing the genus *Shewanella* under saline condition. Degradation by new strain of *Alcaligenes faecalis* of Reactive Orange 13 was studied over the pH range of 5.0–10.0 in nutrient broth, whereas a still wider pH range of 2.0–11.0, using mineral salts medium during degradation of Malachite Green by *Sphingominas paucimobilis*.

4.5 Role of salinity

Common salt is added to dye bath at high concentration to enhance dye fixation. A limitation in the development of bioremediation of textile wastewater is the suceptibility of numerous dye decolorizing bacteria to elevated concentration of sodium chloride. Hence, it is important to decide the effective NaCl cocentrations for the efficient dye decolorization/degradation by bacteria. Salinity in the range of 1–9% at intervals of 2% was checked for textile dye degradation using *Stenotrophomonas maltophila* RSV-2. Khalid *et al.* [69] also examined *Shewanella* sp. for their decolorization/degradation potential in salinity ranging from 0% to 10% (w/v) of NaCl. *Exiguobacterium* sp., a new salt tolerant organism screened from the top soil near a pharmaceutical plant, China was subjected to 2–15%v/v concentration range of NaCl in semisynthetic medium to study its dye degradation. Likewise, acclamatized natural consortium in the form of activated sludge was employed for degradation of dyes by Dafale *et al.* [70] in the range of 0–10% concentration of NaCl for degradation of RB

5. A similar study was conducted by Oturkar *et al.* [19] using 1–5% salt concentration for RR120 decolorization by *B. lentus* BI377.

4.6 Role of Cosubstrates/electron donors

Various organic acids, amino acids, sugars, and organic nitrogenous ingredients act as redox mediators supporting the electron dependent azo bond breakdown by the degrading strain. Therefore, different types of cosubstrates/electron donors need to be enumerated for the decolorizing strain for optimization of the degradation process. However, such studies do not follow a recommended list to outline the cosubstrate/electron donor affecting the degradation kinetics. Yet, many researchers have attempted to investigate the part of cosubstrate/reductant for the azo dye degradation. For instance, various carbon sources such as glucose, mannitol, yeast extract, and maltose at 0.4% concentration were tested for their effect as cosubstrate/electron donor on degradation of Remazol Black-B (100 mg/L) by five bacterial isolates using minimal salt medium [20]. Similarly, Junnarkar *et al.* [37] examined decolorization/degradation by novel bacterial consortium NBNJ6 of Direct Red 81 using the 0.1% of cosubstrates/electron donors such as dextrose, sucrose, cellobiose, and starch in combination with yeast extract. They also used 0.1% starch in combination with 0.1% each of peptone, casein, tryptone, meat extract, and likewise employing starch and casein individually in Bushnell and Hass medium. In the same way, various concentrations of yeast extract (10–50 mg %) as cosubstrates/electron donors were tested by Pandey and Dubey [37] to check for the biodegradation efficiency of *Alcaligenes* sp. AA09 with RR-BL as a model dye. Various carbon sources (0.5–5%) such as glucose, arabinose, fructose, raffinose, rhamnose, xylose, starch, sucrose, and nitrogen nutrients similar to peptone, yeast extract, tryptone, beef extract, soya-bean meal were supplemented in mineral salt medium by Nachiyar and Rajakumar [38] during degradation study of Navitan Fast Blue S5R using *P. aeruginosa*. Similarly, Saratale *et al.* [59] attempted Scarlet R biodegradation with consortium GR in presence of 1% each of nitrogen and/or carbon supplying nutrients such as sucrose, starch, malt extract, lactose, glucose, casein, beef extract, yeast extract, urea, peptone added to the synthetic medium. In an alternative study, Tamboli *et al.* [66] studied the degradation of Direct Red 5B along with the production of Polyhydroxyalkanoates (PHA) by *Sphingobacterium* sp. ATM using basal medium with glycerol, glucose, starch, molasses, and fried oil as different carbon sources and cheese whey and urea as the source of nitrogen. Likewise, the degradation of five dyes with the help of bacterial consortium HM-4 was studied by Khehra *et al.* [71] in mineral salts medium with 0–7.0 mM glucose and 0–0.15% w/v of yeast extract as cosubstrates/electron donors. Similar cosubstrates were used by Moosvi *et al.* [26] in Bushnell Hass medium with 0.05% yeast extract and 0.1% glucose during degradation of RV 5 by the consortium RVM 11.1 wherein the impact of other carbon sources such as combinations of glucose, sucrose, starch, hydrolyzed starch, sodium acetate, lactose, sodium formate with yeast extract was also studied. For the biodegradation of Disperse Red 54 by *B. laterosporus*, Kurade *et al.* [49] used 0.5% w/v of various nutrient sources such as starch, glucose and 0.5% w/v of peptone, yeast extract, urea, and NH_4Cl in Bushnell Hass medium. Eleven types of carbon nutrients such as galactose, xylose, glucose, sucrose, maltose, lactose, raffinose, mannitol, starch, carboxy methyl cellulose glycogen, and four nitrogen nutrients such as YE, peptone, urea, meat extract were investigated for their influences on the decolorization/degradation of four water-soluble dyes using various dye house effluent bacterial isolates [31]. Rajeswari *et al.* [48]

also studied the effect of 0.05%, 0.1%, 0.2%, 0.4%, and 0.8% w/v yeast extract on the textile dyes decolorization/degradation using *Stenotrophomonas maltophila* RSV-2. During the degradation of Red 2G by *Bacillus* sp. isolated from textile effluent, different carbohydrates such as dextrose, sucrose, maltose, starch, cellulose were used in M-9 medium. Also, different nitrogen sources such as peptone, tryptone, tyrosine, glycine, and ammonium ferrous sulfate were checked for their ability to support dye degradation by their isolate [72]. A different set of cosubstrates/electron donors were analyzed by Chen *et al.* [39]. They analyzed the effect of organic acids with one carbohydrate using 20 mM each of lactate, formate, butyrate, pyruvate, acetate, arabinose on Crystal Violet degradation by *Shewanella* species NTOU1.

4.7 Role of alternate electron acceptors

Azo dye degradation is majorly considered as membrane-bound metabolic activity in the absence of oxygen using dye molecule as an oxidant for the re-oxidation of reducing equivalents produced during metabolism. Any other external electron acceptor such as few ions having favorable redox potential acts at the terminal point of electron transport chain and can quench the electrons otherwise used in the breakdown of the dye. For instance, the nitrate and nitrite salts are generally used in dye baths to improve substantivity to the cloth fibers [39, 73, 74]. However, the mechanism used by the decolorizing strain is redox mediated catalysis for breakdown of various bonds present in the synthetic dyes. Nonetheless, in the enzymatic breakdown of the dye, interference by few ions may lower the rate of the degradation process by accepting the electrons that were originally designated for breaking the dye bonds and acting like an electron sink [75]. Liu *et al.* [76] point out to the above phenomenon in the breakdown of Acid Red 27 (AR 27) by *Shewanella oneidensis* MR-1. They used humic acid as the redox mediator for the degradation of AR 27, which led to increase in decolorization efficiency by 15–29%; however, on increasing the concentrations of the salts of nitrates and nitrites, the color removal, which was vetted in the presence of the redox mediator, was revoked. This may be due to the preferential redox potential of nitrate and nitrite ions as compared to the dye. Similarly, Chen *et al.* [39] scrutinized Crystal Violet degradation by *Shewanella* species NTOUI in the presence of ferric citrate, thiosulfate, nitrite ferric oxide, or manganese oxide as electron acceptors in the medium.

4.8 Influence of initial dye concentration

The ability of the decolorizer to withstand increasing initial concentrations of a dye is closely linked to its toxicity tolerance to the dye under higher concentration. Also, the ability to tolerate increasing initial dye concentrations is linked to the structure of the dye. For instance, Oturkar *et al.* [19] subjected *B. lentus* BI377 to 250–1500 ppm of RR120 for checking the maximum decolorization/degradation performance of the bacterium. Similarly, 50 mg/L–1500 mg/L of Crystal Violet was used for studying the degradation by *Shewanella* sp. NTOUI [39]. Jadhav *et al.* [52] checked degradative proficiency of *Comamonas* sp. UVS with dye concentration ranging from 50 ppm–1100 ppm. Similarly, medium amendment was done using 100–500 mg/L at intervals of 50 mg/L of Remazol Black B while studying the dye degradation by *P. putida* [24]. Likewise, Pandey and Dubey [37] studied the effect of 50 ppm–500 ppm of RRBL during the degradation by *Alcaligenes* sp. AA09. While, for the biodegradation of Congo Red using *Bacillus* sp. the dye concentration range was 100 ppm–1000 ppm [32]. Also, Khalid *et al.* [69] have studied the degradation of 100–500 mg/L concentration of AR 88 and Direct Red 81 using *Shewanella putrefaciens* AS96. For the degradation of RV 5 by the consortium named RVM 11.1, the concentration range from 50 ppm to 400 ppm was used [26] while for an acclimatized microbial consortium the initial dye concentration of 200 ppm–1000 ppm of Reactive Black 5 was tested for its degradation ability [70].

4.9 Anaerobic azo dye degradation

Under anaerobic conditions, acidogenic, acetogenic, and methanogenic bacteria have been used for dye decolorization/degradation [77]. The catabolic organic nutrient source is required by these bacteria for dye decolorization/degradation. The nutrients such as acetate, ethanol, starch, glucose, tapioca, and whey were checked during decolorization of dye under methanogenic environments [73, 78–80]. The claim of methanogens being associated with dye decolorization/degradation has also been supported by Razo-Flores *et al.* [81]. In contrast to the above finding Yoo *et al.* [80] confirmed Orange 96 decolorization/degradation in the presence of an inhibitor specific to methanogens, i.e., 2-bromoethanesulfonic acid (BES), and the reduction in degradation was supported by the role of sulfate-reducing bacteria (SRB) in this process. The first step during azo dye degradation under anaerobic and aerobic conditions using bacteria is the reduction of the azo interaction. It may be either extracellular or intracellular. The reduction may take place by separate mechanisms, for example, redox mediators of less molecular weight, chemical reduction by biologically generated reductant similar to sulfide, enzymes, or a blend of all these [82]. The degradation process may be taking place as dye may serve as an oxidant to electrons donated by electron carrying complexes of the Respiratory Electron Transport Chain (RETC). In order to accomplish this, the bacteria should create an interaction between their membrane-bound RETC and the extracellular azo dye molecules. To form such interaction, the RETC components have to be present in the external wall surface of the Gram-negative bacteria in which they can be in direct association with either the substrate dye or the redox mediators present on the outer surface of the cell. Additionally, redox mediator compounds of low molecular weight can represent as electron shuttle for azoreductase dependent on NADH specifically placed in the external wall layer and the azo dye. These mediator compounds may be added externally or are formed by the bacteria from certain substrates during their metabolism. The azo dye reduction without enzymes can take place in presence of very low concentrations of synthetic redox mediators such as anthraquinone sulfonates. However, under aerobic condition, oxygen inhibits this reduction mechanism because of the redox mediator being preferred for oxidation by O_2, instead of the dye [83].

4.10 Aerobic azo dye degradation

In literature, a variety of bacterial species that can reduce azo dyes aerobically have been described. The organic carbon/energy sources are obligatory, as they cannot make use of dye as the only carbon/energy/nitrogen nutrient [84]. *P. aeruginosa* needed glucose while decolorizing Navitan Fast Blue S5R, a commercial textile and tannery dye, under aerobic conditions. It could also decolorize azo dyes range [38]. An incredibly smaller number of bacteria have the ability to use azo dyes as the solitary carbon compound. Such bacterial species abstractly cleave –N=N– and generate amines that they can be used for carbon and energy metabolism. Coughlin *et al.* [85]

have reported an obligate aerobe, *Sphingomonas* sp., strain 1CX, which can use dye AO7, as the only energy and carbon/nitrogen resource. It can degrade only one amine, 1-amino 2-naphthol, along with other amines created throughout AO7 degradation. *Sphingomonas* ICX decolorized quite a few azo dyes having phenyl or naphthyl groups such as 1-amino-2-naphthol/2-amino-1-naphthol.

5. Microbiological dye degradation processes in combination with advanced oxidation processes (AOPs)

To develop a vigorous and profitable option for azo dye elimination, it is promising to combine Advanced Oxidation Processes (AOPs) with microbial methods. There can be improvisation in the advantages and minimization of the disadvantages of each culture by combining them together [86–89]. The aim of pairing AOPs with biological processes is to permit incomplete degradation of the dye molecules by using the AOPs followed by the comparatively low-cost biological process for supplementary removal of the dye [86, 87, 90, 91]. Hence, the key objective of AOPs is to alter recalcitrant compounds like dyes into smaller degradation intermediates, which are further degradable by microbial processes [87, 88, 90, 92] using microbes isolated from municipal plants or the textile effluents [93, 94] or from dye-contaminated soils [90]. Using this strategy, it is possible to reduce COD significantly in a short time [92, 95, 96]. Some dyes, for example, Orange II inhibit bacteria, so pretreatment using AOP is beneficial as toxicity of dye can be eliminated and can be subsequently degraded biologically [95, 97]. The wastewater with dyes can be subjected to such combination of physicochemical and biological methods with an objective of increasing the biodegradability index (BI) due to increase in BOD and decrease in COD [98, 99]. There are also evidences of preliminary decolorization/degradation of the azo dye using microbial methods followed by AOP as the post-treatment method [92, 100, 101]. The studies on the decolorization/degradation of synthetic dyes by combined methods for treatment under various conditions have been enlisted in **Table 1**.

Sonolysis increases biodegradability of dye by decomposing it into smaller units [94]. A better choice is ozonation used in association with biological methods, as the former method augments sludge settling ability and bio-decomposition leading to good color elimination of dyes, as it hits azo bonds even with smaller concentrations [93, 98, 113]. The oxidation using Fenton reagent is not appropriate as a post-treatment application [88]. It is a useful process for getting rid of organic pollutants/toxic chemicals, which are detrimental to biological treatments [86, 94, 114, 115]. It is appropriate to apply Fenton process prior to treatment than after treatment as the Fe ion quantity has to be increased for the complete removal of the dye, since there is iron precipitation due to the phosphate ion, as a macronutrient present in microbiological media. A combination of microbiological processes with electrochemical methods demonstrates superior results as elevated decolorization/degradation with good decrease in COD during the decomposition of dyes [95, 96]. Ultraviolet (UV) light in association with H_2O_2 and TiO_2 as catalyst is commonly employed for the azo dye degradation [112]. The advantages of UV/H_2O_2 process are that it can take place under ambient environment and there is likelihood of total simplification of organic compounds into carbon dioxide [116], low primary investment, absence of objectionable solid leftover, or odor release during or post degradation [117]. Usually, there is a formation of oxygen in this method, which is useful for successive aerobic biological methods [116], but the

S. No.	Biological agent	Enzymes/ sludge/ Chemical agents	Dye/ Effluent	Process Details	AOP*	Process Details	Efficiency	Reference
1	*Consortium of P. aeruginosa, Bacillus flexus* and *Staphylococcus lentus*		Acid Blue 113	Aerobic	Fenton	0.5:1 w/w H_2O_2/COD and 70 mg/L of Fe^{+2}, pH 2.5–3.5	Pre-treatment of Acid Blue 113 dye effluent with Fenton (H_2O_2 and Fe^{2+}) reduced concentration of dye by 40% and 45% biodegradation was achieved by consortium. Overall 89.5% degradation and 93.7% COD reduction was achieved.	[102]
2	*Aeromonas hydrophila* SK16		Reactive Red 180 (RR 180), Reactive Black 5 (RB 5) and Remazol Red (RR)	Aerobic (enzymes detected: tyrosinase, laccase, LiP, ribofavin reductase and azoreductase)	H_2O_2	4% H_2O_2 irradiated under solar light for 6 h. The same procedure was followed for dark conditions. Control was without H_2O_2.	Bio–AOP led to 100% decolorization of RR180, RB5 and Remazol Red, while 72% decolorization was achieved with individual treatments. Combined treatment reduced BOD and COD of RR 180 by 78 and 68%, RB 5 by 52 and 83% and RR by 42 and 47%, respectively	[89]

S. No.	Biological agent	Enzymes/ sludge/ Chemical agents	Dye/ Effluent	Process Details	AOP*	Process Details	Efficiency	Reference
3	*Consortium of Chaetomium globosum* IMA1, *Aspergillus niger and Rhizopus oryzae*		polyvinyl alcohol and organic compounds in dye effluents	Aerobic (Enzymes detected: laccase and LiP)	Fenton reaction with H_2O_2	Temperature: 25 ± 2 °C. Chemicals: Ferrous sulfate heptahydrate and hydrogen peroxide (H_2O_2)	97.8%, 98.5% and 99.75% COD, Polyvinyl alcohol and color removal respectively using combined method. Complete degradation was observed after 6 days.	[103]
4	*Consortium of Brevibacillus laterosporus and Galactomyces geotrichum*		Raw Textile effluent (RTE)—1 and 2, and Simulated Synthetic Effluent (SSE)	Reproducibility of the bioreactor maintained/3 consecutive cycles of 7 days each	Continuous degradation	Triple layered fixed bed reactor, >80% decolorization/ degradation/100 mL/h flow-rate, for 7 days, 78% COD.	Decolorization efficiency–89% for RTE-1, 60% for RTE-2 and 69% for SSE in 48 h.	[104]
5	*Lactobacillus* sp., *Mycobacterium* sp., *Staphylcoccus aureus*, and *Corynbacterium* sp. from Distillery wastewaters		Textile effluent	Aerobic and microaerophilic conditions	Electro-Fenton (EF) process	Textile wastewater pH 3.0 for Electro-Fenton method, pH 12.5 for precipitation Fe^{3+} ions and pH 6.6 for biological treatment	Removal of color, COD, TOC by EF process was 63%, 48%, 29% respectively. Further treating wastewater by biological process reduced 85% color, 86% COD, and 56% of TOC	[105]

S. No.	Biological agent	Enzymes/ sludge/ Chemical agents	Dye/ Effluent	Process Details	AOP*	Process Details	Efficiency	Reference
6	*Aeromonas veronii* GRI		Methyl orange	Aerobic under neutral and alkaline pH	Biosurfactant	0.075% SPB-1 lipopeptide from *Bacillus subtilis* SPB-1	92% degradation in SPB-1 biosurfactant–microbial treatment	[106]
7		Laccase extracted from *Cyathus bulleri*, procured from culture collection	Textile effluent	Pretreatment with ABTS and laccase 60% decolorization	Chemical treatment with ABTS and Alum	Alum coagulated 90% of residual color but the process associated with dye sludge formation and ABTS not recovered	85% degradation was achieved in the Enzyme Membrane Reactor and the process could be operated for over a period of 15 days. No sludge formation, negligible membrane fouling. About 60% ABTS could be recovered.	[107]
8	Not mentioned	Not mentioned	Dyeing wastewaters	Biological oxidation 25°C/12 h; 1 h feeding/ reaction for 6 h/ sedimentation for 4 h/discharge for 0.8 h/idle for 0.2 h/cycles-10; pH ~7.0	Fenton's oxidation followed with Sequential biological oxidation	Oxidation using Fenton reagent, with optimum dosages of H_2O_2 and Fe (II)	91–98% for DOC, 83–95% for BOD5 and 88–98% for COD, color removal >99% compared to biological or chemical treatment alone. Lowered toxicity by inhibition of *Vibrio fischeri*.	[108]

S. No.	Biological agent	Enzymes/ sludge/ Chemical agents	Dye/ Effluent	Process Details	AOP*	Process Details	Efficiency	Reference
9	Consortium of strains of *P. aeruginosa* from different sites of petroleum pipelines		Turquoise Blue G	Aerobic	Electrolysis	IrO_2–RuO_2–TiO_2–Ti anode; Ti cathode 5 g/L NaCl, pH 7.0 current density 10 mA/cm^2	Complete decolorization of Turquoise Blue G within 15 min. Dye components were reduced by 41.3% by electrochemical and 90% by microbial route.	[109]
10		Activated sludge from wastewater treatment plant with initial microbial concentration of 0.5 g/L	Methylene Blue	Aerobic–Activated sludge	Electrochemical	25 mA/cm^2 Pb/PbO_2 Electrode.	95.61% degradation efficiency was achieved for 134 mg/L methylene Blue at 720 rpm. Biodegradability from 0.034 to 0.54 for the BOD_5/COD ratio with electrolysis and 92.03% mineralization was achieved	[110]
11	Consortium of *A. niger and Penicillium* sp. from tannery yard		Acid Black 1	280 h/Aerobic	Ozonation	Ozonation 4 h	52.9% TOC and 94.5% color removal	[99]

S. No.	Biological agent	Enzymes/ sludge/ Chemical agents	Dye/ Effluent	Process Details	AOP*	Process Details	Efficiency	Reference
12	Consortium of *Pleurotus ostreatus, Tricoderma viridae* or *Fusarium oxysporum Bacillus cereus and Pseudomonas putida*		Procion Blue	RT/150 rpm Aerobic/5 days.	Electrochemical	Electrochemical reactor, 4 h. Photo catalysis 500 mg/L TiO_2, 5 h–125 Hg lamp, Electrolysis 5.0 A/ dm^2, 5 h	2000 mg/L of Procion Blue, reduction in COD—39.6% in electrochemical reactor; COD—31–59% post-treatment with microbes; COD–32–81% on electrolysis and COD–23–70% after photocatalytic process.	[96]
13	*P. putida, B. cereus, Phanerochaete chrysosporium, Tricoderma virdae*		Methyl Red	Biological at 25°C for 5 days	Electro-oxidation	Electro-oxidized RuO_2; stainless steel cathode and titanium anode coated with–2A/dm^2, 5 h.	77–91% COD and 81% color removal post electro-oxidation followed with biological 90–95% color removal	[95]
14		Anaerobic sludge/ Aerobic sludge. (municipal effluent plant)	Remazol Red	Anaerobic process at pH 7.0 for 14 days	Photocatalysis and Fenton reagent	Aerobic, pH 7.0, 150 rpm; UV–Vis lamp 18 W, 1 mM Ferrous, 1 mMolar H_2O_2, pH 5.0 and 50 rpm.	100 ppm of Remazol Red–COD 70 ppm for anaerobic; COD 14 ppm post Photo-Fenton treatment; Post-treatment of aerobic conditions—100% color removal; residual COD 9 mg/L, less toxic	[100]

DOI: http://dx.doi.org/10.5772/intechopen.107913

S. No.	Biological agent	Enzymes/ sludge/ Chemical agents	Dye/ Effluent	Process Details	AOP*	Process Details	Efficiency	Reference
15	*Rhodotorula mucilaginosa*		Basic Yellow 2 and Reactive blue 4 and Reactive red 2	100 rpm/25°C/ pH 5/8 h	Ultrasound	Ultrasonication 20 kHz, 5 h	Ultrasonication of 100 ppm dyes removed 28–48% color. Post-microbial treatment color removal was 40–93%.	[94]
16		Hydrogen peroxide Photo-oxidation	Reactive blue 5	H_2O_2 250–1000 ppm, pH 7.0, RT, 21 W low-pressure Hg lamp λ_{254nm}.	Ultraviolet	UV 15–60 min	RB5 (200 ppm), UV H_2O_2 89–100% color; 54% decrease in COD, high biodegradibility	[90]
17		Aerobic sludge	Remazol Black B	over activated carbon, pH 9.0/24 h	Ozonation	Ozonation RT, 60 min, pH 3.0.	96% color removal of 100–500 ppm dye, high toxicity with Ozonation, not toxic with biological post-treatment.	[93]
18		Aerobic treatment	Reactive Brilliant Red X-3B	Microbiological aerated filter, 20–25°C	Ozonation	Ozonation dye/ ozone = 4.5, pH 11.0.	50 mg/L dye showed 30% COD reduction with ozonation. Post biological treatment, 97% color removal and 90% decrease in COD was observed.	[26]

S. No.	Biological agent	Enzymes/ sludge/ Chemical agents	Dye/ Effluent	Process Details	AOP*	Process Details	Efficiency	Reference
19		Biological reactor	Cibacron Red	Immobilized sludge, 40°C,with oxygen saturation	Photo-Fenton	Photo-Fenton 65–225 ppm H_2O_2 2–5 ppm Fe (II)	250 mg/L dye showed 41–67% decrease in COD on photo-Fenton treatment while 60–73% was achieved on biodegradation	[92]
20	*Candida oleophila*	*Candida oleophila*	Reactive Blue 5	Aerobic, 26°C	Fenton oxidation	Fenton H_2O_2 1.5×10^{-3} mol/L, ferrous ion 1.5×10^{-4} mol/L at pH 5.0	100–500 ppm dye showed 35% and 15% color removal on treatment with Fenton's method and *Candida* sp. However, 1 h preatreatment with Fenton followed by microbial treatment for 7 days resulted in 91–95% reduction in color.	[111]
21	*Trametes versicolor*	*Trametes versicolor*	Reactive Blue 5	Immobilization 25°C, 4 days	Photocatalysis Ultraviolet TiO_2	254 nm with 15 W high pressure mercury-vapor lamp, at 120 rpm	98% color of 300 ppm dye after microbial or combined H_2O_2/UV/TiO_2 treatment. Poisonous to *H. atenuatta*; only photocatalysis–non-toxic.	[112]

AOP: advanced oxidation processes.

Table 1.
Comparison of various combined methods under various conditions.

higher UV light intensity required for the better rate of cleavage of dyes contributes to the disadvantage because the process utilizes extensive electricity, which increases the operating cost [118]. Hence the blend of ultraviolet light/H_2O_2 and biological methods can reduce the cost. For instance, the degradation of Reactive Black 5, which is partially treated with photochemical step and then by treatment with an acclimatized microbial biomass is rapid, hence consumes lesser H_2O_2 concentrations [90].

6. Generalized mechanisms of azo dye degradation

6.1 Enzyme catalyzed dye degradation

Rafii *et al.* [119] produced the first report of azoreductases in bacteria growing anaerobically. The bacteria belonging to the genera *Eubacterium* and *Clostridium* displayed decolorization of azo dyes of sulfonated type while growing on chemically undefined solid media. These strains produced extracellular oxygen-sensitive azoreductases constitutively. This enzyme from *Clostridium perfringens* alleged to be flavoprotein dehydrogenase and was found to be occupied in the reduction of nitro aromatic compounds. Another mechanism involving flavin nucleotide dependent cytosolic reductases, which donate electrons through soluble flavins to dyes, has also been suggested for dye decolorization/degradation [116]. However, Russ *et al.* [120] used *Sphingomonas* strain-BN6 produced by recombinant technology and showed the insignificance of the above mechanism in vivo.

6.1.1 Enzymes involved in dye decolorization/degradation

The progression of biodegradation of azo dye occurs due to the presence of a variety of oxidoreductases present in microorganisms.

The two important classes of azoreductases with respect to their catalysis are azoreductases of flavoprotein kind and those that do not need flavin. The former class of azoreductase is further subdivided on the basis of their requirement for coenzymes such as NADH, NADPH, or both, as reductants [121]. Other reductases such as riboflavin reductase, NADH-DCIP reductase might be implicated in the azo dye cleavage [46, 122, 123]. However, as per Blumel *et al.* [124] and Russ *et al.* [120], these enzymes are futile in vivo. Russ *et al.* [120] also proposed that the anaerobic azoreductases of cytoplasmic origin are flavin reductases and can engage themselves in extracellular hydrogenation of azo dyes with the help of an electron mediator, which helps in displacement of reducing counterparts to azo dyes from the membrane of bacteria. Thus, the presence of azoreductases along with a transport system that enables dye transport to the cells is a requirement for the microbial strains to decolorize azo dyes.

6.1.1.1 Reductive enzymes

A large number of different types of bacteria have been screened and identified, which can degrade azo dyes under oxygen deficient (reduced) environment [19]. It is agreed that an enzyme azoreductase initially cleaves the azo bond in such conditions resulting in creation of colorless aromatic amines. The microaerophilic environment augments azo dye degradation as revealed by findings conceded by Joshi *et al.* [125] and Xu *et al.* [51]. Reactive azo dyes can be degraded rapidly under partly reduced environment. To create partly reduced environment, the presence of another redox agent for transfer of electrons from reduced nicotinamide coenzyme to the dyes is essential as observed by Chang *et al.* [126]. Though a small fraction of azo dyes can be degraded aerobically [127], the efficiency of degradation may decline if the oxygen is present [128], since oxygen and the dyes compete with each other as electron acceptors. Aerobic conditions enhance cell mass of microorganisms but with less efficiency of degradation process in comparison with reduced environment. These observations lead to the interpretation that it is the concentration of oxygen and not the amount of cell mass that influences the process of dye degradation. Likewise, the process is subject to the nature and accessibility of electron donors [31, 129]. Both cellular components, namely membrane fractions isolated from bacteria and cytoplasm, are important during anaerobic reduction process [130]. But most significant is the membrane fractions because they harbor constituents for electron conveyance which link azo dyes to electron donors. This concludes that the biochemical basis for azo dye decomposition is the redox reaction between azo dyes and electron donors [131]. The most important reductases are azoreductase, riboflavin reductase, and DCIP reductase [46, 122, 123].

Anaerobic degradation of azo dye is a nonspecific procedure. It was understood that in the absence of O_2, the azo dye is degraded by azoreductase by means of reducing equivalents such as NADH or NADPH. The primary stage during decolorization of dye by bacteria is the –N=N– bond reduction. This is carried out by transferring four electrons across the azo bond with the help of NADH. The dye serves as the terminal electron acceptor instead of O_2 under static/anoxic condition and two sequential reduction steps reduce the same to specific amines [126, 132]. For illustration, Misal *et al.* [133] had procured alkaliphilic bacterium *Bacillus badius* for dye degradation from alkaline Crater Lake–Lonar; subsequently, purifiying and characterizing its azoreductase. Likewise, many authors have reviewed the role of the azoreductase enzyme in azo dye decolorization/degradation. However, a variation of soluble cytoplasmic azoreductases with small substrate specificity are produced by bacteria [59], e.g., azoreductases from certain anaerobic bacteria such as *Clostridium* and *Eubacterium,* which are oxygen-sensitive, are produced constitutively, and excreted into the ecosystem [119]. Nachiyar and Rajakumar [134] purified oxygen-resistant intracellular azoreductase from *Pseudomonas aeruginosa* and studied its affinity for various azo dyes pointing out its highest affinity for Navitan Fast Blue S5R.

6.1.1.2 Oxidative enzymes

The enzymes of oxidative kind, which are important in the dye degradation, include tyrosinase, lignin peroxidase (LiP), laccase, and manganese peroxidase (MnP) [135]. As compared to bacterial isolates, white-rot fungi has shown tremendous potential in degradation of dyes and other recalcitrant compounds by production of highly oxidative and substrate nonspecific enzymes [35]. Among the oxidative enzymes, LiP and MnP are extensively studied for their potential to degrade azo dyes. Both these enzymes are multi-copper phenol oxidases and have wider applications during oxidation of a variety of partially degraded by-products of dye. Also, unlike bacterial enzymes, fungal LiP and MnP follow a highly nonspecific free radical mechanism during azo dye degradation to form phenolic compounds without cleaving the azo bond. In this process, they skip the biochemical steps that lead to formation of toxic aromatic amines [20, 36, 136–138]. Telke *et al.* [46] further reported an enzyme, phenol oxidase, similar to laccase having the characteristic

S. No.	Enzyme	Microbial system	Dye decolorized	Reference
1	Crude protease	*Bacillus cereus* strain KM201428	Reactive Black 5	[142]
2	Laccase like enzyme Lac 1326	Marine metagenomic library. Cloned and overexpressed in *Escherichia coli* BL21	Amaranth, Coomassie Brilliant Blue, Bromophenol Blue, Acid Violet 7, Congo Red, and Indigo Carmine	[143]
3	Laccase, tyrosinase, LiP, Riboflavin reductase, azoreductase	*Aeromonas hydrophila* SK16 and AOPs	RR180, Reactive Black 5 and Remazol Red	[89]
4	Laccase, LiP	*A. hydrophila*	Crystal Violet	[144]
5	Laccase, LiP, MnP	*Peyronellaea prosopidis*	Scarlet RR	[145]
6	Laccase, NADH-DCIP reductase, tyrosinase, LiP, Malachite Green reductase	*Aeromonas* sp. DH-6	Malachite Green	[47]
7	Laccase, azoreductase	*A. hydrophila* SK16 and *Lysinibacillus sphaericus* SK13	Reactive Yellow F3R, Joyfix Yellow 53R, Remazol Red RR, Drimaren Black CL-S and Disperse Red F3BS	[34]
8	Laccase, MnP, LiP, azoreductase	*Anoxybacillus flavithermus* 52-1A, *Tepidiphilus thermophilus* JHK30, *Tepidiphilus succinatimandens* 4BON, *Brevibacillus aydinogluensis* PDF25, *Bacillus thermoamylovorans* DKP, *Geobacillus thermoleovorans* NP1	Direct Black G	[146]
9	Laccase, LiP, tyrosinase, Riboflavin reductase	*Ipomoea hederifolia, Cladosporium cladosporioides* (Plant-Fungus consortium)	Navy Blue HE2R	[147]
10	Azoreductase	*Shewanella* sp. strain IFN4	Acid Red-88, RB5, DR81	[148]
11	Laccase and MnP	*Leptosphaerulina* sp. (Fungus)	Novacron Red	[149]
12	Laccase, azoreductase, Veratryl alcohol oxidase, NADH-DCIP reductase	*Pseudomonas* species SUK1 and *Providencia rettgeri* HSL1 in consortium	C.I. RO 16, C.I. RB 5, C.I. DR 81 and C.I. Disperse Red 78	[11]
13	NADH–DCIP reductase, LiP	*Penicillium simplicissimum*	Triphenylmethane Dyes	[150]
14	Laccase	*Trichoderma atroviride* F03	RB5	[151]

S. No.	Enzyme	Microbial system	Dye decolorized	Reference
15	Laccase	*Stenotrophomonas maltophilia* AAP56	RB5	[152]
16	Laccase	*Coprinopsis cineria*	Methyl Orange	[153]
17	Azoreductase	*Bacillus lentus* BI377	RR141	[154]
18	Azoreductase	*Shewanella oneidensis* MR-1	Methyl Red	[155]
19	Laccase, LiP, MnP	*Ganoderma lucidum* IBL-05	RR195a Reactive Yellow 145a Reactive Blue 21	[156]
20	Veratryl Alcohol Oxidase	*Alcaligenes faecalis* PMS-1	Reactive Orange 13	[157]
21	NADH/NADPH-dependent O_2 sensitive azoreductase	*Alcaligenes* sp. AA09	RR-BL	[37]
21	Laccase, azoreductase, and NADH–DCIP reductase	*Shewanella aquimarina*	Acid Red 27, Direct Blue 71, RR120, Methyl Orange, Acid Orange7	[158]
22	Tyrosinase	*Brevibacterium* sp.	Direct Blue 71, RY107, RB 5, RR 198	[56]
23	Veratryl alcohol oxidase	*Pseudomonas aeruginosa* Strain BCH	Remazol Black	[159]
24	Heat stable laccase	*Bacillus pumilus*	Acetosyringone, Indigocarmine	[160]
25	Laccase, NADH–DCIP reductase, tyrosinase, LiP	*Acinetobacter calcoaceticus* NCIM 2890	Amaranth Dye	[161]
26	Azoreductase, Cytochrome P450 oxidase, Aminopyrine N-demethylase, Superoxide dismutase, Glutathione S-transferase, tyrosinase	*B. lentus* BI377	RR120	[19]
27	NADH dependent O_2 insensitive azoreductase	*Bacillus badius*	Amaranth Dye	[133]
28	Azoreductase	Moderately Halotolerant *Bacillus megaterium*	Red 2G	[138]

Table 2.
Dye degradation using microbial enzymes.

to use non-phenolic substrates. Laccases can employ direct oxidation or mediator coupled indirect oxidation of textile dyes by H2O2 during their catalytic cycle [139]. On the other hand, MnP activity is dependent on manganese as well as specific buffers. For this reason, the use of enzymatic membrane reactors is suggested by some authors for degradation of dyes using MnP. It was reported that peroxidase involved in dye degradation can degrade hydroxyl free anthraquinone dyes [140, 141]. Few examples of studies employing biodegradation with the help of oxidative and reductive enzymes are enlisted in **Table 2**.

6.2 Mediated biological azo dye decolorization/degradation

Many azo dyes, which have a large molecular weight or are polymeric azo dyes, or are strongly polar sulfonate, are difficult to be imported via the cytoplasmic membrane [162]. The recommendation to this was that there could be another mechanism for reduction of these dyes. There are currently various reports discussing the function of redox mediators during anoxic bacterial reduction of dyes [163, 164]. The addition of catalytic quantities of riboflavin to anaerobic granular sludge led to appreciable increase of the decolorization of mordant yellow 10 [165]. Mendez-Paz *et al.* [166] introduced 1-amino-2-napthol, an amine derived during AO7 degradation that improved its rate of reduction, probably through the acquisition of electrons. The presence of artificial electron carrier like anthraquinone-2, 6-disulfonate similarly enhanced the reduction of numerous dyes. Keck *et al.* [164] have observed that the augmentation of dye decolorization, under anaerobic conditions, took place when coupled with redox intermediates produced by other bacteria while degrading aromatic compounds aerobically. With respect to this observation, Chang *et al.* [132] too showed improved azo dye decolorization/degradation rates post addition of cell-free supernatants having metabolic intermediates of a dye-decolorization by *E. coli* strain NO3.

7. Conclusion

Microorganisms play an important role in decomposition and mineralization of synthetic dyes. The microbial processes are, in turn, affected by various environmental, molecular, and physicochemical factors. The biodegradation of dye molecules is also dependent on multiple factors including structure and stability (types of bonds) of dye molecules. More importantly, dyes are present in industrial effluents as a mixture of recalcitrant compounds, organic and chemical compounds. Many of these compounds interfere in biodegradation of individual dyes resulting in impractical outcomes on field application in spite of successful laboratory studies. Although several researchers have presented detailed mechanisms of microbial strains to potentially decolorize a number of synthetic dyes, it is important to understand that dye degradation is a complicated multifactorial and multistep process. The interference of pollutants severely compromises biodegradation either by inhibition of microorganisms or their enzyme activity. To make matters much more challenging, factors such as temperature, dissolved oxygen, moisture, and coexistence of dyes with acid, alkali, and other pollutants further impede biodegradation process. In such a scenario, in situ augmentation of natural dye degrading microorganisms may be a simple, practical, and sustainable bioremediation strategy. However, even for this purpose, it is essential to screen natural isolates capable of degrading complex molecular structures and identify optimum parameters to maximize the degrading potential of screened isolates. These studies can provide significant insights into the specific role of enzymes, mediators, cofactors, and cosubstrates that will ensure augmentation of beneficial dye degrading strains. In addition, these studies can ensure selection of most effective combination of physical techniques to minimize toxicity and enhance biodegradation potential of microorganisms. Collectively, these strategies can detoxify and completely degrade complex mixtures of dye molecules into harmless metabolites.

Author details

Radhika Birmole and Aruna K. Samudravijay*
Department of Microbiology, Wilson College, Mumbai, Maharashtra, India

*Address all correspondence to: arunasam2000@gmail.com

References

[1] Desore A, Narula SA. An overview on corporate response towards sustainability issues in textile industry. Environment Development and Sustainability. 2018;**20**(4):1439-1459

[2] Synthetic Dye Global Market Report 2022—By Type (Aniline Dyes, Chrome Dyes, Anionic Dyes, Cationic Dyes), By Product Type (Acid Dyes, Disperse Dyes, Reactive Dyes, Direct Dyes, Basic Dyes, VAT Dyes, Other Product Types), By End-User Industry (Textile, Food & Beverages, Paper, Ink, Leather, Other End-Use Industries)—Market Size, Trends, And Global Forecast 2022-2026. Available from: https://www.thebusinessresearchcompany.com/report/synthetic-dye-global-market-report#:~:text=The%20global%20synthetic%20dyes%20market,a%20CAGR%20of%2010.7%25. [Accessed: August 31, 2022]

[3] Berradi M, Hsissou R, Khudhair M, Assouag M, Cherkaoui O, El Bachiri A, et al. Textile finishing dyes and their impact on aquatic environs. Heliyon. 2019;**5**(11):e02711. DOI: 10.1016/j.heliyon.2019.e02711

[4] Chequer FMD, Dorta DJD, de Oliveira DP. Azo dyes and their metabolites: Does the discharge of the azo dye into water bodies represent human and ecological risks? In: Hauser PJ, editor. Advances in Treating Textile Effluent [Internet]. London: IntechOpen; 2011. DOI: 10.5772/19872. Available from: https://www.intechopen.com/chapters/22392. [Accessed: August 31, 2022]

[5] Lellis B, Fávaro-Polonio CZ, Pamphile JA, Polonio JC. Effects of textile dyes on health and the environment and bioremediation potential of living organisms. Biotechnology Research and Innovation. 2019;**3**(2):275-290

[6] Piaskowski K, Świderska-Dąbrowska R, Zarzycki PK. Dye removal from water and wastewater using various physical, chemical, and biological processes. Journal of AOAC International. 2018;**101**(5):1371-1384

[7] Isidori M, Lavorgna M, Nardelli A, Parrella A, Previtera L, Rubino M. Ecotoxicity of naproxen and its phototransformation products. The Science of the Total Environment. 2005;**348**(1-3):93-101

[8] Wang W, Nykamp J, Huang XD, Gerhardt K, Dixon DG, Greenberg BM. Examination of the mechanism of phenanthrenequinone toxicity to *Vibrio fischeri*: Evidence for a reactive oxygen species—Mediated toxicity mechanism. Environmental Toxicology and Chemistry. 2009;**28**:1655-1662

[9] Jamee R, Siddique R. Biodegradation of synthetic dyes of textile effluent by microorganisms: An environmentally and economically sustainable approach. European Journal of Microbiology & Immunology. 2019;**9**(4):114-118

[10] Ledakowicz S, Paździor K. Recent achievements in dyes removal focused on advanced oxidation processes integrated with biological methods. Molecules (Basel, Switzerland). 2021;**26**(4):870

[11] Lade H, Kadam A, Paul D, Govindwar S. Biodegradation and detoxification of textile azo dyes by bacterial consortium under sequential microaerophilic/aerobic processes. EXCLI Journal. 2015;**14**:158-174

[12] Ngo A, Tischler D. Microbial degradation of azo dyes: Approaches and

prospects for a hazard-free conversion by microorganisms. International Journal of Environmental Research and Public Health. 2022;**19**(8):4740

[13] Qiu H, Shen F, Yin A, Liu J, Wu B, Li Y, et al. Biodegradation and detoxification of azo dyes by halophilic/halotolerant microflora isolated from the salt fields of tibet autonomous region China. Frontiers in Microbiology. 2022;**13**:877151

[14] Mohanty SS, Kumar A. Enhanced degradation of anthraquinone dyes by microbial monoculture and developed consortium through the production of specific enzymes. Scientific Reports. 2021;**11**(1):7678

[15] Maier J, Kandelbauer A, Erlacher A, Cavaco-Paulo A, Gübitz GM. A new alkali-thermostable azoreductase from *Bacillus* sp. strain SF. Applied and Environmental Microbiology. 2004;**70**(2):837-844

[16] Cong J, Xie X, Liu Y, Qin Y, Fan J, Fang Y, et al. Biochemical characterization of a novel azo reductase named BVU5 from the bacterial flora DDMZ1: Application for decolorization of azo dyes. RSC Advances. 2022;**12**(4):1968-1981

[17] Ponraj M, Gokila ZV. Bacterial decolorization of textile dye (Orange 3R). International Journal of Advanced Biotechnology and Research. 2011;**2**(1):168-177

[18] Ponraj M, Jamunarani P, Zambare V. Isolation and optimization of conditions for decolorization of true blue by textile dye decolorizing fungi. Asian Journal of Experimental and Biological Sciences. 2011;**2**(2):270-277

[19] Oturkar CC, Nemade HN, Mulik PM, Patole MS, Hawaldar RR, Gawai KR. Mechanistic investigation of decolorization and degradation of Reactive Red 120 by *Bacillus lentus* BI377. Bioresource Technology. 2011;**102**(2):758-764

[20] Mahmood S, Arshad M, Khalid A, Nazli ZH, Mahmood T. Isolation and screening of azo dye decolorizing bacterial isolates from dye-contaminated textile wastewater. Soil and Environment. 2011;**30**(1):7-12

[21] Usha MS, Sasirekha B, Bela RB, Devi S, Kamalini C, Manasa GA, et al. Batch, repeated batch and continuous degradation of Reactive Black 5 and Reactive Red 120 dye by immobilized bacteria. Journal of Scientific and Industrial Research. 2012;**71**(7):504-510

[22] Guadie A, Tizazu S, Melese M, Guo W, Hao H. Biodecolorization of textile azo dye using *Bacillus* sp. strain CH12 isolated from alkaline lake. Biotechnology Reports. 2017;**15**:92-100

[23] Junnarkar N, Murty DS, Bhatt NS, Madamwar D. Decolorization of diazo dye Direct Red 81 by a novel bacterial consortium. World Journal of Microbiology and Biotechnology. 2006;**22**(2):163-168

[24] Kannan S, Dhandayuthapani K, Sultana M. Decolorization and degradation of azo dye-Remazol Black B by newly isolated *Pseudomonas putida*. International Journal of Current Microbiological Application. 2013;**2**(4):108-116

[25] Khalid A, Arshad M, Crowley DE. Accelerated decolorization of structurally different azo dyes by newly isolated bacterial strains. Applied Microbiology and Biotechnology. 2008;**78**(2):361-369

[26] Moosvi S, Keharia H, Madamwar D. Decolourization of textile dye Reactive

Violet 5 by a newly isolated bacterial consortium RVM 11.1. World Journal of Microbiology and Biotechnology. 2005;**21**(5):667-672

[27] Nigam P, Banat IM, Singh D, Marchant R. Microbial process for the decolorization of textile effluent containing azo, diazo and reactive dyes. Process Biochemistry. 1996;**31**(5):435-442

[28] Sharma DK, Saini HS, Singh M, Chimni SS, Chadha BS. Biological treatment of textile dye Acid Violet-17 by bacterial consortium in an up-flow immobilized cell bioreactor. Letters in Applied Microbiology. 2004;**38**(5):345-350

[29] Fang H, Wenrong H, Yuezhong L. Investigation of isolation and immobilization of a microbial consortium for decoloring of azo dye 4BS. Water Research. 2004;**38**(16):3596-3604

[30] Birmole R, Patade S, Sirwaiya V. Biodegradation study of Reactive Blue 172 by *Shewanella haliotis* DW01 isolated from lake sediment. Indian Journal of Scientific Research. 2014;**5**(1):139-152

[31] Modi HA, Rajput G, Ambasana C. Decolorization of water soluble azo dyes by bacterial cultures, isolated from dye house effluent. Bioresource Technology. 2010;**101**(16):6580-6583

[32] Gopinath KP, Sahib AMH, Muthukumar K, Velan M. Improved biodegradation of Congo Red by using *Bacillus* sp. Bioresource Technology. 2009;**100**(2):670-675

[33] Supaka N, Juntongjin K, Damronglerd S, Delia ML, Strehaiano P. Microbial decolorization of reactive azo dyes in a sequential anaerobic-aerobic system. Chemical Engineering Journal. 2004;**99**(2):169-176

[34] Srinivasan GP, Sikkanthar A, Elamaran A, Delma CR, Subramaniyan K, ThirugnanasambandanSS.Biodegradation of carcinogenic textile azo dyes using bacterial isolates of mangrove sediment. Journal of Coastal Life Medicine. 2014;**2**(2):154-162

[35] Niladevi KN, Prema P. Effect of inducers and process parameters on laccase production by *Streptomyces psammoticus* and its application in dye decolourization. Bioresource Technology. 2008;**99**:4583-4589

[36] Mohana S, Shrivastava S, Divecha J, Madamwar D. Response surface methodology for optimization of medium for decolorization of textile dye Direct Black 22 by a novel bacterial consortium. Bioresource Technology. 2008;**99**(3):562-569

[37] Pandey AK, Dubey V. Biodegradation of azo dye Reactive Red BL by *Alcaligenes* sp. AA09. International Journal of Engineering and Science. 2012;**1**(12):54-60

[38] Nachiyar CV, Rajakumar GS. Degradation of a tannery and textile dye, Navitan Fast Blue S5R by *Pseudomonas aeruginosa*. World Journal of Microbiology and Biotechnology. 2003;**19**(6):609-614

[39] Chen KC, Wu JY, Liou DJ, Hwang SCJ. Decolorization of the textile dyes by newly isolated bacterial strains. Journal of Biotechnology. 2003;**101**(1):57-68

[40] Khan J. Biodegradation of azo dye by moderately halotolerant *Bacillus megaterium* and study of enzyme azoreductase involved in

degradation. Advances in Biotechnology. 2011;**10**(7):21-27

[41] Sugiura W, Miyashita T, Yokoyama T, Arai M. Isolation of azo-dye-degrading microorganisms and their application to white discharge printing of fabric. Journal of Bioscience and Bioengineering. 1999;**88**(5):577-581

[42] Alhassani HA, Rauf MA, Ashraf SS. Efficient microbial degradation of Toluidine Blue dye by *Brevibacillus* sp. Dyes and Pigments. 2007;**75**(2):395-400

[43] Su WT, Lin CH. Fungal-bacterial synergism enhanced decolorization of Reactive Red 120 by response surface methodology. International Biodeterioration and Biodegradation. 2013;**82**:1-8

[44] Nawahwi MZ, Ibrahim Z, Yahya A. Degradation of the azo dye Reactive Red 195 by *Paenibacillus* spp. R2. Bioremediation and Biodegradation. 2013;**4**(1):1-7

[45] Eichlerova I, Homolka L, Nerud F. Decolorization of high concentrations of synthetic dyes by the white rot fungus *Bjerkandera adusta* strain CCBAS 232. Dyes and Pigments. 2007;**75**(1):38-44

[46] Telke AA, Kalyani DC, Dawkar VV, Govindwar SP. Influence of organic and inorganic compounds on oxidoreductive decolorization of sulfonated azo dye C. I Reactive Orange 16. Journal of Hazardous Materials. 2009;**172**(1):298-309

[47] Du LN, Li G, Zhao YH, Xu HK, Wang Y, Zhou Y, et al. Efficient metabolism of the azo dye Methyl Orange by *Aeromonas* sp. strain DH-6: Characteristics and partial mechanism. International Biodeterioration and Biodegradation. 2015;**105**:66-72

[48] Rajeswari K, Subashkumar R, Vijayaraman K. Decolorization and degradation of textile dyes by *Stenotrophomonas maltophilia* RSV-2. International Journal of Environmental Bioremediation and Biodegradation. 2013;**1**(2):60-65

[49] Kurade MB, Waghmode TR, Khandare RV, Jeon BH, Govindwar SP. Biodegradation and detoxification of textile dye Disperse Red 54 by *Brevibacillus laterosporus* and determination of its metabolic fate. Journal of Bioscience and Bioengineering. 2016;**121**(4):442-449

[50] Pillai HPJS. Optimization of process conditions for effective degradation of Azo Blue dye by *Streptomyces* DJP15. Journal of Pure and Applied Microbiology. 2017;**11**(4):1757-1765

[51] Xu M, Guo J, Sun G. Biodegradation of textile azo dye by *Shewanella decolorationis* S12 under microaerophilic conditions. Applied Microbiology and Biotechnology. 2007;**76**(3):719-726

[52] Jadhav JP, Parshetti GK, Kalme SD, Govindwar SP. Decolourization of azo dye Methyl Red by *Saccharomyces cerevisiae* MTCC 463. Chemosphere. 2007;**68**(2):394-400

[53] Ben Mansour H, Mosrati R, Corroler D, Ghedira K, Barillier D, Chekir L. In vitro mutagenicity of Acid Violet 7 and its degradation products by *Pseudomonas putida* mt-2: Correlation with chemical structures. Environmental Toxicology and Pharmacology. 2009;**27**(2):231-236

[54] Elisangela F, Andrea Z, Fabio DG, de Menezes CR, Regina DL, Artur CP. Biodegradation of textile azo dyes by a facultative *Staphylococcus arlettae* strain VN-11 using a sequential microaerophilic/aerobic process. International Biodeterioration and Biodegradation. 2009;**63**(3):280-288

[55] Kalme SD, Parshetti GK, Jadhav SU, Govindwar SP. Biodegradation of benzidine based dye Direct Blue-6 by *Pseudomonas desmolyticum* NCIM 2112. Bioresource Technology. 2007;**98**(7):1405-1410

[56] Elisangela F, Grossman MJ, Paschoal JA, Reyes FG, Durrant LR. Decolorization and biodegradation of reactive sulfonated azo dyes by a newly isolated *Brevibacterium* sp. strain VN-15. Springer Plus. 2012;**1**(1):37-43

[57] Lin J, Zhang X, Li Z, Lei L. Biodegradation of Reactive Blue 13 in a two-stage anaerobic/aerobic fluidized beds system with a *Pseudomonas* sp. isolate. Bioresource Technology. 2010;**101**(1):34-40

[58] Pourbabaee AA, Malekzadeh F, Sarbolouki MN, Najafi F. Aerobic decolorization and detoxification of a disperse dye in textile effluent by a new isolate of *Bacillus* sp. Biotechnology and Bioengineering. 2006;**93**(4):631-635

[59] Saratale RG, Saratale GD, Chang JS, Govindwar SP. Ecofriendly degradation of sulfonated diazo dye C.I. Reactive Green 19A using *Micrococcus glutamicus* NCIM-2168. Bioresource Technology. 2009;**100**(17):3897-3905

[60] Jadhav SU, Kalme SD, Govindwar SP. Biodegradation of Methyl Red by *Galactomyces geotrichum* MTCC 1360. International Biodeterioration and Biodegradation. 2008;**62**(2):135-142

[61] Parshetti GK, Telke AA, Kalyani DC, Govindwar SP. Decolorization and detoxification of sulfonated azo dye Methyl Orange by *Kocuria rosea* MTCC 1532. Journal of Hazardous Materials. 2010;**176**(1-3):503-509

[62] Ren S, Guo J, Zeng G, Sun G. Decolorization of triphenylmethane, azo, and anthraquinone dyes by a newly isolated *Aeromonas hydrophila* strain. Applied Microbiology and Biotechnology. 2006;**72**(6):1316-1321

[63] Chen CH, Chang CF, Ho CH, Tsai TL, Liu SM. Biodegradation of Crystal Violet by a *Shewanella* sp. NTOU1. Chemosphere. 2008;**72**(11):1712-1720

[64] Saratale RG, Saratale GD, Chang JS, Govindwar SP. Bacterial decolorization and degradation of azo dyes: A review. Journal of the Taiwan Institute of Chemical Engineers. 2011;**42**(1):138-157

[65] Jadhav UU, Dawkar VV, Ghodake GS, Govindwar SP. Biodegradation of Direct Red 5B, a textile dye by newly isolated Comamonas sp. UVS. Journal of Hazardous Materials. 2008;**158**(2-3):507-516

[66] Tamboli DP, Kagalkar AN, Jadhav MU, Jadhav JP, Govindwar SP. Production of polyhydroxyhexadecanoic acid by using waste biomass of *Sphingobacterium* sp. ATM generated after degradation of textile dye Direct Red 5B. Bioresource Technology. 2010;**101**(7):2421-2427

[67] Saratale RG, Saratale GD, Kalyani DC, Chang JS, Govindwar SP. Enhanced decolorization and biodegradation of textile azo dye Scarlet R by using developed microbial consortium-GR. Bioresource Technology. 2009;**100**(9):2493-2500

[68] Li T, Guthrie JT. Colour removal from aqueous solutions of metal-complex azo dyes using bacterial cells of *Shewanella* strain J18 143. Bioresource Technology. 2010;**101**(12):4291-4295

[69] Khalid A, Arshad M, Crowley DE. Decolorization of azo dyes by *Shewanella spunder* saline conditions. Applied

Microbiology and Biotechnology. 2008;**79**(6):1053-1059

[70] Dafale N, Rao NN, Meshram SU, Wate SR. Decolorization of azo dyes and simulated dye bath wastewater using acclimatized microbial consortium–biostimulation and halotolerance. Bioresource Technology. 2008;**99**(7):2552-2558

[71] Khehra MS, Saini HS, Sharma DK, Chadha BS, Chimni SS. Decolorization of various azo dyes by bacterial consortium. Dyes and Pigments. 2005;**67**(1):55-61

[72] Pushpa V, Yogendra K, Mahadevan M, Kalasaiah M, Aroonsrimorakot S. Effect of carbon and nitrogen sources for the degradation of red 2G by *Bacillus* sp. International Journal of Pharmaceutical Sciences Review and Research. 2017;**47**(21):108-113

[73] Carliell CM, Barclay SJ, Naidoo N, Buckley CA, Mulholland DA, Senior E. Microbial decolourisation of a reactive azo dye under anaerobic conditions. Water Sabinet African. 1995;**21**(1):61-69

[74] Clark M. Handbook of Textile and Industrial Dyeing: Principles, Processes and Types of Dyes. Cambridge: Woodhead Publishing, Elsevier; 2011

[75] Dubin P, Wright KL. Reduction of azo food dyes in cultures of *Proteus vulgaris*. Xenobiotica. 1975;**5**(9):563-571

[76] Liu G, Zhou J, Wang J, Wang X, Jin R, Lv H. Decolorization of azo dyes by *Shewanella oneidensis* MR-1 in the presence of humic acids. Applied Microbiology and Biotechnology. 2011;**91**(2):417-424

[77] Kasper HF, Wuhrmann K. Kinetic parameters and relative turnover of some important catabolic reactions in digesting sludge. Applied and Environmental Microbiology. 1978;**36**(1):1-7

[78] Chinwetkitvanich S, Tuntoolvest M, Panswad T. Anaerobic decolorization of reactive dyebath effluents by a two stage UASB system with tapioca as cosubstrate. Water Research. 2000;**34**:2223-2232

[79] Talarposhti AM, Donnelly T, Anderson GK. Colour removal from a simulated dye wastewater using a two phase anaerobic packed bed reactor. Water Research. 2001;**35**:425-432

[80] Yoo ES, Libra J, Adrian L. Mechanism of decolorization of azo dyes in an anaerobic mixed culture. Journal of Environment Engineering (ASCE). 2001;**127**(9):844-849

[81] Razo-Flores E, Luijten M, Donlon B, Lettinga G, Field J. Biodegradation of selected azo dyes under methanogenic conditions. Water Science and Technology. 1997;**36**:65-72

[82] Pandey A, Singh P, Iyengar L. Bacterial decolorization and degradation of azo dyes. International Biodeterioration and Biodegradation. 2007;**59**(2):73-84

[83] Pearce CI, Lloyd JR, Guthrie JT. The removal of colour from textile wastewater using whole bacterial cells: A review. Dyes and Pigments. 2003;**58**(3):179-196

[84] Stolz A. Basic and applied aspects in the microbial degradation of azo dyes. Applied Microbiology and Biotechnology. 2001;**56**(1-2):69-80

[85] Coughlin MF, Kinkle BK, Bishop PL. Degradation of azo dyes containing aminonaphthol by *Sphingomonas* sp. strain 1CX. Journal of Industrial Microbiology and Biotechnology. 1999;**23**(4-5):341-346

[86] Mohajerani M, Mehrvar M, Ein-Mozaffari F. An overview of the integration of advanced oxidation technologies and other processes for water and waste- water treatment. International Journal Engineering. 2009;**3**:120-146

[87] Oller I, Malato S, Ssnchez-Perez JA. Combination of advanced oxidation processes and biological treatments for wastewater decontamination—A review. Science of the Total Environment. 2011;**409**:4141-4166

[88] Prato-Garcia D, Buitron G. Degradation of azo dye mixtures through sequential hybrid systems: Evaluation of three advanced oxidation processes for the pre-treatment stage. Journal of Photochemistry Photobiology A: Chemistry. 2011;**223**:103-110

[89] Thanavel M, Kadam SK, Biradar SP, Govindwar SP, Hun B. Combined biological and advanced oxidation process for decolorization of textile dyes. Springer Nature Applied Sciences. 2019;**1**(97):1-16

[90] Dokuzoglu Z, Alkan U. Biotreatability enhancement of aqueous Reactive Black 5 by hydrogen peroxide/ultraviolet advanced oxidation process. Color Technology. 2010;**126**:308-314

[91] Holkar CR, Jadhav AJ, Pinjari DV, Mahamuni NM, Pandit AB. A critical review on textile wastewater treatments: Possible approachesJournal of. Environmental Management. 2016;**182**:351-366

[92] Garcia-Montano J, Torrades F, Perez-Estrada LA, Oller I, Malato S, Maldonado MI, et al. Degradation pathway of the commercial reactive azo dye Procion Red H-E7B under solar assisted Photo-Fenton reaction. Environment Science and Technology. 2008;**42**:6663-6670

[93] deSouza U, Bonilla KAS, de Souza AAU. Removal of COD and color from hydrolyzed textile azo dye by combined ozonation and biological treatment. Journal of Hazardous Materials. 2010;**179**:35-42

[94] Onat TA, Gumus HT, Guvenc A, Donmez G, Mehmetoglu U. Decolorization of textile azo dyes by ultrasonication and microbial removal. Desalination. 2010;**255**:154-158

[95] Babu BR, Parande AK, Kumar SA, Bhanu SU. Treatment of dye effluent by electrochemical and biological processes. Open Journal of Safety Science and Technology. 2011;**1**:12-18

[96] Basha CA, Selvakumar KV, Prabhu HJ, Sivashanmugam P, Lee CW. Degradation studies for textile reactive dye by combined electrochemical, microbial and photocatalytic methods. Separation and Purification Technology. 2011;**79**(3):303-309

[97] Babu BR, Parande AK, Raghu S, Kumar TP. An overview of wastes produced during cotton textile processing and effluent treatment methods. Journal of Cotton Sciences. 2007;**11**:110-122

[98] Baban A, Yediler A, Avaz G, Hostede SS. Biological and oxidative treatment of cotton textile dye-bath effluents by fixed and fluidized bed reactors. Bioresource Technology. 2010;**101**(4):1147-1152

[99] Kanagaraj J, Mandal AB. Combined biodegradation and ozonation for removal of tannins and dyes for the reduction of pollution loads. Environmental Science and Pollution Research. 2012;**19**:42-52

[100] Jonstrup M, Punzi M, Mattiasson B. Comparison of anaerobic pre-treatment and aerobic post-treatment coupled to photo-Fenton oxidation for degradation of azo dyes. Journal of Photochemistry and Photobiology A: Chemistry. 2011;**224**(1):55-61

[101] Kim S, Park C, Kim TH, Lee J, Kim SW. COD reduction and decolorization of textile effluent using a combined process. Journal of Bioscience and Bioengineering. 2003;**95**(1):102-105

[102] Shanmugam BK, Easwaran SN, Mohanakrishnan AS, Kalyanaraman C, Mahadevan S. Biodegradation of tannery dye effluent using Fenton's reagent and bacterial consortium: A biocalorimetric investigation. Journal of Environmental Management. 2019;**242**:106-113

[103] Ajmi K, Vismara E, Manai I, Haddad M, Hamdi M, Bouallagui H. Polyvinyl acetate processing wastewater treatment using combined Fenton's reagent and fungal consortium: Application of central composite design for conditions optimization. Journal of Hazardous Materials. 2018;**358**:243-255

[104] Kurade MB, Waghmode TR, Patil SM, Jeon BH, Govindwar SP. Monitoring the gradual biodegradation of dyes in a simulated textile effluent and development of a novel triple layered fixed bed reactor using a bacterium-yeast consortium. Chemical Engineering Journal. 2017;**1**(307):1026-1036

[105] Roshini PS, Gandhimathi R, Ramesh ST, Nidheesh PV. Combined electro-Fenton and biological processes for the treatment of industrial textile effluent: Mineralization and toxicity analysis. Journal of Hazardous, Toxic, and Radioactive Waste. 2017;**21**(4):1-8

[106] Mnif I, Maktouf S, Fendri R, Kriaa M, Ellouze S, Ghribi D. Improvement of Methyl Orange dye biotreatment by a novel isolated strain, *Aeromonas veronii* GRI, by SPB1 biosurfactant addition. Environmental Science and Pollution Research. 2016;**23**(2):1742-1754

[107] Chhabra M, Mishra S, Sreekrishnan TR. Combination of chemical and enzymatic treatment for efficient decolorization/degradation of textile effluent: High operational stability of the continuous process. Biochemical Engineering Journal. 2015;**93**:17-24

[108] Rodrigues CSD, Madeira LM, Boaventura RAR. Synthetic textile dyeing wastewater treatment by integration of advanced oxidation and biological processes–performance analysis with costs reduction. Journal of Environmental Chemical Engineering. 2014;**2**:1027-1039

[109] Pramila M, Manikandan S, Anju KS, Murali Kannan M, Hong S, Maruthamuthu S, et al. Electrochemical decolorization and degradation of Turquoise Blue G (TBG) by pre-adapted petroleum degrading bacteria. Separation and Purification Technology. 2014;**132**:719-727

[110] Yahiaoui I, Aissani-Benissad F, Madi K, Benmehdi N, Fourcade F, Amrane A. Electrochemical pre-treatment combined with biological treatment for the degradation of Methylene Blue dye: Pb/PbO_2 electrode and modeling-optimization through central composite design. Industrial and Engineering Chemistry Research. 2013;**52**(42):14743-14751

[111] Lucas MS, Dias AA, Sampaio A, Amaral C, Peres JA. Degradation of a textile reactive azo dye by a combined chemical–biological process: Fenton's reagent-yeast. Water Research. 2007;**41**:1103-1109

[112] El-Dein AM, Libra JA, Wiesmann U. Kinetics of decolorization and mineralization of the azo dye Reactive Black 5 by hydrogen peroxide and UV light. Water Science Technology. 2001;**44**:295-301

[113] Khare UK, Bose P, Vankar PS. Impact of ozonation on subsequent treatment of azo dye solutions. Journal of Chemical Technology & Biotechnology. 2007;**82**(11):1012-1022

[114] Elias B, Guihard L, Nicolas S, Fourcade F, Amrane A. Effect of electro-Fenton application on azo dyes biodegradability. Environmental Progress and Sustainable Energy. 2011;**30**:160-167

[115] Hilal NM. Treatment of reactive dyeing wastewater by different advanced oxidation processes. Der Chemica Sinica. 2011;**2**:262-273

[116] Rafii F, Cerniglia CE. Comparison of the Azoreductase and Nitroreductase from *Clostridium perfringens*. Applied Environmental Microbiology. 1993;**59**(6):1731-1734

[117] Aleboyeh A, Olya ME, Aleboyeh H. Electrical energy determination for an azo dye decolorization and mineralization by UV/H_2O_2 advanced oxidation process. Chemical Engineering Journal. 2008;**137**:518-524

[118] Blanco J, Torrades F, Varga M, Garcia-Montano J. Fenton and biological-Fenton coupled processes for textile wastewater treatment and reuse. Desalination. 2011;**286**:394-399

[119] Rafii F, Franklin W, Cerniglia CE. Azoreductase activity of anaerobic bacteria isolated from human intestinal microflora. Applied and Environmental Microbiology. 1990;**56**:2146-2151

[120] Russ R, Rau J, Stolz A. The function of cytoplasmic flavin reductases in the reduction of azo dyes by bacteria. Applied and Environmental Microbiology. 2000;**66**(4):1429-1434

[121] Chen H, Hopper SL, Cerniglia CE. Biochemical and molecular characterization of an azoreductase from *Staphylococcus aureus*, a tetrameric NADPH-dependent flavoprotein. Microbiology. 2005;**151**(5):1433-1441

[122] Ghodake G, Jadhav S, Dawkar V, Govindwar S. Biodegradation of diazo dye Direct Brown MR by *Acinetobacter calcoaceticus* NCIM 2890. International Biodeterioration and Biodegradation. 2009;**63**(4):433-439

[123] Kalyani DC, Telke AA, Dhanve RS, Jadhav JP. Ecofriendly biodegradation and detoxification of Reactive Red 2 textile dye by newly isolated *Pseudomonas* sp. SUK1. Journal of Hazardous Materials. 2009;**163**(2-3):735-742

[124] Blumel S, Knackmuss HJ, Stolz A. Molecular cloning and characterization of the gene coding for the aerobic azoreductase from *Xenophilus azovorans* KF46F. Applied Microbiology and Biotechnology. 2002;**68**:3948-3955

[125] Joshi T, Iyengar L, Singh K, Garg S. Isolation, identification and application of novel bacterial consortium TJ-1 for the decolourization of structurally different azo dyes. Bioresource Technologies. 2008;**99**:7115-7121

[126] Chang JS, Chou C, Lin YC, Lin PJ, Ho JY, Hu TL. Kinetic characteristics of bacterial azo-dye decolorization by *Pseudomonas luteola*. Water Research. 2001;**35**(12):2841-2850

[127] Sarayu K, Sandhya S. Aerobic biodegradation pathway for Remazol Orange by *Pseudomonas aeruginosa*.

Applied Biochemistry and Biotechnology. 2010;**160**:1241-1253

[128] Pearce CI, Christie R, Boothman C, von Canstein H, Guthrie JT, Lloyd JR. Reactive azo dye reduction by *Shewanella* J18 143. Biotechnology and Bioengineering. 2006;**95**(4):692-703

[129] Brige A, Motte B, Borloo J, Buysschaert G, Devreese B, Van Beeumen JJ. Bacterial decolorization of textile dyes is an extracellular process requiring a multicomponent electron transfer pathway. Microbial Biotechnology. 2008;**1**(1):40-52

[130] Kudlich M, Keck A, Klein J, Stolz A. Localization of the enzyme system involved in anaerobic reduction of azo dyes by *Sphingomonas* sp. strain BN6 and effect of artificial redox mediators on the rate of azo dye reduction. Applied. Environmental Microbiology. 1997;**63**(9):3691-3694

[131] Mahmood S, Khalid A, Arshad M, Mahmood T, Crowley DE. Detoxification of azo dyes by bacterial oxidoreductase enzymes. Critical Reviews in Biotechnology. 2015;**36**(4):639-651

[132] Chang J, Chen B, Lin YS. Stimulation of bacterial decolorization of an azo dye by extracellular metabolites from *Escherichia coli* strain NO3. Bioresource Technology. 2004;**91**:243-248

[133] Misal SA, Lingojwar DP, Shinde RM, Gawai KR. Purification and characterization of azoreductase from alkaliphilic strain *Bacillus badius*. Process Biochemistry. 2011;**46**(6):1264-1269

[134] Nachiyar CV, Rajakumar GS. Purification and characterization of an oxygen insensitive azoreductase from *Pseudomonas aeruginosa*. Enzyme and Microbial Technology. 2005;**36**(4):503-509

[135] Aftab U, Khan MR, Mehfooz M, Ali M, Aslam SH, Rehman A. Decolourization and degradation of textile azo dyes by *Corynebacterium* sp. isolated from industrial effluent. Pakistan. Journal of Zoology. 2011;**43**(1):1-8

[136] Kurniawati S, Nicell JA. Efficacy of mediators for enhancing the laccase-catalyzed oxidation of aqueous phenol. Enzyme and Microbial Technology. 2007;**41**(3):353-361

[137] Kurniawati S, Nicell JA. A comprehensive kinetic model of laccase-catalyzed oxidation of aqueous phenol. Biotechnology Progress. 2009;**25**:763-773

[138] Morozova OV, Shumakovich GP, Gorbacheva MA, Shleev SV, Yaropolov AI. Blue laccases. Biochemistry. 2007;**72**:1136-1150

[139] Khlifi R, Belbahri L, Woodward S, Ellouz M, Dhouib A, Sayadi S, et al. Decolourization and detoxification of textile industry wastewater by the laccase-mediator system. Journal of Hazardous Materials. 2010;**175**:802-808

[140] Marchis T, Avetta P, Bianco-Prevot A, Fabbri D, Viscardi G, Laurenti E. Oxidative degradation of Remazol Turquoise Blue G 133 by soybean peroxidase. Journal of Inorganic Biochemistry. 2011;**105**(2):321-327

[141] Sugano Y, Matsushima Y, Shoda M. Complete decolorization of the anthraquinone dye Reactive Blue 5 by the concerted action of two peroxidises from *Thanatephorus cucumeris* Dec1. Applied Microbiology and Biotechnology. 2006;**73**:862-871

[142] Wanyonyi WC, Onyari JM, Shiundu PM, Mulaa FJ. Effective biotransformation of Reactive Black 5 dye using crude protease from *Bacillus*

cereus Strain KM201428. Energy Procedia. 2019;**157**:815-824

[143] Boran F, Birhanli E, Yesilada O, Ozbey E. Comparison of indigo carmine decolorization by *Pseudomonas aeruginosa* and crude laccase enzyme from *Funalia trogii*. Turkish Journal of Biology. 2019;**43**:37-46

[144] Bhargava R, Mani S, Mulla S, Saratale G. Degradation and decolourization potential of an ligninolytic enzyme producing *Aeromonas hydrophila* for crystal violet dye and its phytotoxicity evaluation. Ecotoxicology and Environmental Safety. 2018;**156**:166-175

[145] Bankole PO, Adekunle AA, Obidi OF, Chandanshive VV, Govindwar SP. Biodegradation and detoxification of Scarlet RR dye by a newly isolated filamentous fungus, *Peyronellaea prosopidis*. Sustainable Environment Research. 2018;**28**(5):214-222

[146] Chen Y, Feng L, Li H, Wang Y, Chen G, Zhang Q. Biodegradation and detoxification of Direct Black G textile dye by a newly isolated thermophilic microflora. Bioresource Technology. 2017;**250**:650-657

[147] Patil SM, Chandanshive VV, Rane NR, Khandare RV, Watharkar AD, Govindwar SP. Bioreactor with Ipomoea hederifolia adventitious roots and its endophyte *Cladosporium cladosporioides* for textile dye degradation. Environmental Research. 2016;**146**:340-349

[148] Imran M, Arshad M, Negm F, Khalid A, Shaharoona B, Hussain S, et al. Yeast extract promotes decolorization of azo dyes by stimulating azoreductase activity in *Shewanella* sp. strain IFN4. Ecotoxicology and Environmental Safety. 2016;**124**:42-49

[149] Placido J, Chanaga X, Ortiz-Monsalve S, Yepes M, Mora A. Degradation and detoxification of synthetic dyes and textile industry effluents by newly isolated Leptosphaerulina sp. from Colombia. Bioresource Bioprocess. 2016;**3**(1):6-12

[150] Chen SH, Ting ASY. Biosorption and biodegradation potential of triphenylmethane dyes by newly discovered *Penicillium simplicissimum* isolated from indoor wastewater sample. International Biodeterioration & Biodegradation. 2015;**103**:1-7

[151] Adnan LA, Sathishkumar P, Yusoff ARM, Hadibarata T. Metabolites characterisation of laccase mediated Reactive Black 5 biodegradation by fast growing ascomycete fungus *Trichoderma atroviride* F03. International Biodeterioration and Biodegradation. 2015;**104**:274-282

[152] Galai S, Korri-Youssouf H, Nejib MM. Characterization of yellow bacterial laccase SmLac/role of redox mediators in azo dye decolorization. Journal of Chemical Technology and Biotechnology. 2013;**89**(11):1741-1750

[153] Tian YS, Xu H, Peng RH, Yao QH, Wang RT. Heterologous expression and characterization of laccase 2 from *Coprinopsis cinerea* capable of decolourizing different recalcitrant dyes. Biotechnology and Biotechnological Equipment. 2014;**28**(2):248-258

[154] Oturkar CC, Patole MS, Gawai KR, Madamwar D. Enzyme based cleavage strategy of *Bacillus lentus* BI377 in response to metabolism of azoic recalcitrant. Bioresource Technology. 2013;**130**:360-365

[155] Yang Y, Lu L, Gao F, Zhao Y. Characterization of an efficient catalytic and organic solvent-tolerant

azoreductase toward methyl red from *Shewanella oneidensis* MR-1. Environmental Science and Pollution Research. 2014;**20**(5):3232-3239

[156] Sadaf S, Bhatti HN, Bibi I. Efficient removal of disperse dye by mixed culture of *Ganoderma lucidum* and *Coriolus versicolor*. Pakistan Journal of Agricultural Sciences. 2013;**50**:261-266

[157] Shah P, Dave S, Rao M. Enzymatic degradation of textile dye Reactive Orange 13 by newly isolated bacterial strain *Alcaligenes faecalis* PMS-1. International Biodeterioration and Biodegradation. 2012;**69**:41-50

[158] Meng X, Liu G, Zhou J, Shiang FQ, Wang G. Azo dye decolorization by *Shewanella aquimarina* under saline conditions. Bioresource Technology. 2012;**14**:95-101

[159] Phugare SS, Kalyani DC, Patil AV, Jadhav JP. Textile dye degradation by bacterial consortium and subsequent toxicological analysis of dye and dye metabolites using cytotoxicity, genotoxicity and oxidative stress studies. Journal of Hazardous Materials. 2011;**186**(1):713-723

[160] Reiss R, Ihssen J, Thony-Meyer L. *Bacillus pumilus* laccase: A heat stable enzyme with a wide substrate spectrum. BMC Biotechnology. 2011;**11**(9):1-11

[161] Ghodake G, Jadhav U, Tamboli D, Kagalkar A, Govindwar S. Decolorization of textile dyes and degradation of mono-azo dye amaranth by *Acinetobacter calcoaceticus* NCIM 2890. Industrial Journal of Microbiology. 2011;**51**:501-508

[162] Levine WG. Metabolism of azo dyes: Implication for detoxication and activation. Drug Metabolism Reviews. 1991;**23**(3-4):253-309

[163] dos Santos AB, Cervantes FJ, Yaya-Beas RE. Effect of redox mediator, AQDS, on the decolorisation of a reactive azo dye containing triazine group in a thremophilic anaerobic EGSB reactor. Enzyme and Microbial Technology. 2003;**33**(7):942-951

[164] Keck A, Klein J, Kudlich M, Stolz A, Knackmuss HJ, Mattes R. Reduction of azo dyes by redox mediators originating in the naphthalene sulfonic acid degradation pathway of *Sphingomonas* sp. strain BN6. Applied and Environmental Microbiology. 1997;**63**(9):3684-3690

[165] Field JA, Brady J. Riboflavin as a redox mediator accelerating the reduction of the azo dye Mordant Yellow 10 by anaerobic granular sludge. Water Science Technology. 2003;**48**:187-193

[166] Mendez-Paz D, Omil F, Lema JM. Anaerobic treatment of azo dye acid Orange 7 under fed-batch and continuous conditions. Water Research. 2005;**39**:771-778

Chapter 8

Aromatic Plants: Alternatives for Management of Crop Pathogens and Ideal Candidates for Phytoremediation of Contaminated Land

Maria Banda, Alexis Munyengabe and Wilma Augustyn

Abstract

Crop diseases due to fungal pathogens cause significant resulting economic losses in agriculture. For management of crop diseases, farmers use synthetic pesticides. However, the frequent application of these chemicals leads to accumulation in soil and therefore presenting pollution problems. Essential oils (EOs) sourced from aromatic plants are safer alternatives and are effective against a variety of crops pathogens. In addition to their role as the sources of EOs, aromatic plants are gaining much attention in rehabilitation strategies. In phytoremediation processes, suitable plants species are used to clean-up polluted sites. Mining activities and electricity generation processes have resulted in significant amounts of tailings and coal fly ash. Mine tailings and coal fly ash are disposed in dumpsites, converting productive lands to unusable waste sites. These solid waste materials contain toxic metals and therefore posing serious risks to the health of the environment. Aromatic plants can be cultivated in contaminated sites and therefore be used for restoration of polluted lands. The EOs can be sourced from these aromatic plants as they are free from metal-toxicity and can therefore be used to generate revenues. This review highlights the role of aromatic plants in the control of crops pathogens and also their application in phytoremediation processes.

Keywords: aromatic plants, essential oils, crop diseases, phytoremediation strategies, contaminated sites, toxic metals

1. Introduction

Plant diseases caused by infectious pathogens are a global concern to agriculture, significantly impacting on food security and human health [1]. To control disease outbreaks, synthetic pesticides are applied at regular intervals, but their use is ineffective and associated with health risks to humans and the environment. The use of pesticides leads to build-up of toxicants, which should be replaced by biodegradable alternatives. Pathogens readily acquire resistance to fungicides, making them ineffective crop protection agents. Aromatic plants produce essential oils that are effective for prevention and protection of crops against infectious diseases in the field as well as during storage. In addition to their potential as biopesticides, aromatic plants have gained a lot of interest in phytoremediation strategies. Mining activities produce mine waste, while generation of electricity from coal combustion results in production of fly ash and these solid waste materials are disposed in ash and tailings dams [2]. Harsh conditions prevailing on fly ash and mine tailings dams include unfavorable pH, very low levels of nutrients required for plant growth, and unacceptable concentrations of toxic metals [3]. All these converting valuable lands to unproductive waste disposal sites. Aromatic plants have demonstrated potential in remediation of areas contaminated with toxic levels of heavy metals [4]. These species are high value crops as they produce EOs that are free from metal contamination and therefore can be used simultaneously for restoration strategies and as sources of valuable products.

2. Crop diseases and excessive use of agrochemicals

The infestation of crops leads to a reduction in yield, increased production costs and may even be harmful to human and animal health. Pesticides are effective in decreasing pathogen load. Synthetic pesticides are applied at regular intervals throughout the growing season of the crop. Their use has significantly increased agricultural yields and contributed to improved food quality. However, using synthetic chemicals for crop protection is associated with health risks to humans and the environment. Fungal and bacterial pathogens readily acquire resistance to bactericides and fungicides, making them ineffective crop protection agents. For pesticides to be effective, there is a need to use higher doses or their substitution with new and sometimes, highly toxic products. In many cases the pathogen will become resistant to the pesticide used because of the single compound nature of most pesticides. Global warming and accompanying climate changes have resulted in increased resistance of several bacterial and fungal pathogens [5].

The excessive use and misuse of agrochemicals has led to increased bioaccumulation of toxic metals in soils and water and eventual toxicity to humans via food intake and the environment. Pesticides can remain in soil for long after they are applied, continuing to harm the ecosystem. How the pesticide is bound by soil components, how readily it is degraded and environmental conditions determine how long it can remain in soil [6]. Unacceptable levels of organochlorine pesticides (OCPs) residues and potentially toxic metals (Pb, Cr, Zn, Cu, and Fe) were reported in beans and cowpea [7]. The OCPs are volatile, stable and can be bound to the soil components and persist in air, negatively impacting the health of humans, animals and the environment [8]. Furthermore, in several countries, the level of pesticide residues detected in the medicinal plants were above the permissible limits as prescribed by the World Health Organization [9, 10]. Soil contains an abundance of biologically diverse organisms that are essential for agricultural sustainability and the use of pesticides have contributed to their decline [11]. Azole fungicides are extensively used for control of fungal diseases, but their over-use of azole has resulted to contamination of air, soil and crops, mainly because of their lipophilic characteristic [12]. The long-term use of copper-based pesticides has led to build-up in soil worldwide and presenting a potential public health problem due to Cu entering the food chain [13, 14]. Public pressure is also increasing to reduce the use of synthetic chemical products [15, 16]. Furthermore, the European Union has prohibited the use of contaminant plant protection products since 2020. These drawbacks emphasize the discovery of sustainable and environment-friendly pathogen control practices to manage diseases and ensure the safety of consumers.

3. Plants-based products as alternatives for control of crops pathogens

Crops diseases remain as one of the greatest threats to the sustainable development of society, leading to significant agricultural loss and costs incurred on awareness and the development of management strategies. Recently attention has been given to organic farming and food safety owing to the many challenges of using synthetic chemicals and consequently protection of consumers health. A natural solution to this problem could be the use of essential oils that have been shown to demonstrate antibacterial, fungicidal, herbicidal, nematocidal, acaricidal and insecticidal, properties. EOs are mixtures of volatile bioactive compounds that are obtained from various plant parts. They are characterized by a mixture of secondary metabolites including alcohols, phenolics, aldehydes, ketones, terpenoids and other secondary metabolites contributing synergistically or by additive effect to treat infectious diseases [15, 17].

Fungal pathogens are the major causes of economic losses in agriculture worldwide. The fungal species including *Aspergillus*, *Fusarium*, *Penicillium*, *Phytophthora* and *Botrytis* are among the pathogens that contribute significantly to agricultural losses as they can cause decay, accelerated ripening, and production of mycotoxins [18, 19]. The *Aspergillus* species have contributed to significant losses and the most reported species include *Aspergillus fumigatus*, *Aspergillus flavus* and *Aspergillus niger* which produce aflatoxins that exhibit carcinogenic and mutagenic properties, therefore affecting human health [20, 21]. Azoles are the most widely used for control of diseases caused by Aspergillus genus, but studies demonstrated an increasing prevalence of azole-resistant strains such as *A. fumigatus* and these problems are associated with higher clinical burdens and mortality rates [22, 23]. *Aspergillus fumigatus* is a mold found in soil, compost and releases volatile spores onto air, continuously breathed by humans and could be the cause of one of the most common fungal ailments [24, 25]. Fusarium species cause significant agricultural losses in crops including potato, pea, bean, wheat, corn, cabbage, cucumber and rice worldwide [26, 27]. These pathogens are found in soil, plant, air and aquatic environment causing diseases in humans, animals and plants [28]. *Fusarium* pathogens produce mycotoxins that affect the quality of crop produces and threatening human health [29]. *Fusarium* species infect both plants and humans and are resistant toward antifungal agents including amphotericin B, itraconazole, fluconazole and echinocandins and therefore continues to be a problem for patients with compromised immune systems [30]. *Botrytis cinerea* causes pre- and postharvest decay of various crops such as strawberries and this pathogen is resistant to fungicides including benzimidazole and dicarboximide [31, 32]. *Phytophthora* infestans causes late blight diseases in potato and tomato crops worldwide, affecting the economy and the quality and quantity of the crop [33]. Development of resistance was reported for *Phytophthora* infestans against fluazinam which was attributed to the widespread use of the fungicide [34]. *Penicillium* spp. are the most important cause of postharvest decays of fruits and vegetables, causing blue and green mold [12, 35, 36]. These fungal pathogens contribute to losses in crops such as apple, pear, and citrus fruits and contribute to mycotoxin accumulation in processed fruit products [37, 38]. *Penicillium* species are resistant toward fungicide resistance including thiabendazole, guazatine, imazalil and propiconazole [39, 40].

4. Effectiveness of EOs in the protection against plant pathogens

A variety of plants pathogens that affect agricultural produce include viruses, bacteria, fungi, nematode and parasitic plant. EOs are increasingly recognized as potential pesticides in agriculture. EOs can be used as lures for detecting and monitoring insects. *Coriandrum sativum* and *Nerium indicum* EOs are strong attractants to *Cyrtorhinus lividipennis*, a rice planthopper. Insecticidal activity of EOs is connected to a decrease in acetylcholinesterase activity [19]. There are approximately 350 bacteria known to be phytopathogenic, such as Proteobacteria, Actinobacteria and Firmicite [19]. Nematodes are a very destructive group of plant pathogens and the mode of action of EOs against nematodes include GABA, acetylcholinesterase inhibition and octopamine synapses [16]. Approximately 30% of all crop diseases are as a result of infections by phytopathogenic fungi that can produce toxins and carcinogenic substances [19]. EOs diminish the influence of fungal infections by acting on cell walls, cell wall alterations and modifications to gene expression. Bioactivity depends on the composition of the oil, the functional groups present in the major compounds as well as their synergistic effects. In most cases, the oils are more effective as antimicrobial and anti-insecticidal agents than the major components on their own, indicating the important synergistic contribution of minor compounds in bioactivity.

The EOs of *thyme* and *manuka* demonstrated strong fungistatic activity against *Aspergillus niger*, *Fusarium culmorum*, *Phytophthora cactorum*, demonstrating complete inhibition [41]. *Cymbopogon schoenanthus*, *Lippia multifliora* and *Ocimum americanum* EOs were found to reduce the contamination rates of Colletotrichum dematium and Fusarium spp., Cladosporium sp. and Macrophomina phaseolina [42]. The EO of *Salvia sclarea* and *Salvia dolomitica* demonstrated fungicidal activity against *Aspergillus*, *Penicillium*, *Trichoderma viride* and *Fusarium* species [43]. An added benefit from the application of EO was revealed from a study on Tuta absoluta, a tomato pinworm. Treatment with extracts of *Achillea millefolium* and *Achillea sativum*, reduced the number of infested leaves, which was accompanied by induced release of herbivory plant volatiles as the plant defense mechanism [44]. Numerous studies report the wide use of *Mentha* species for management of plant pathogens and insect pests due to the action of alcohols, phenolics, aldehydes, ketones, terpenoids and other secondary metabolites [45, 46]. The combination of EOs has also demonstrated synergistic effect. For example, the combination of tea tree oil and mint had improved activity against *Aspergillus niger* when mixed together [47]. The combination of more than one EO contributes to the inability of pathogens to develop resistance against the EOs. EOs are promising biocontrol agents because several bioactive compounds contribute to the activity of the oil therefore overcoming pathogen resistance. These plant-based products offer advantages including easy degradation, wide spectrum biological activities, cost-effective, renewable in nature, and demonstrate low toxicity [48]. Cultivation of bioactive EO producing plants is a sustainable method to obtain natural products for purpose as biopesticides. Most EOs produced can be used in a variety of applications from industrial to agriculture and in many cases the plants can be cultivated using environmentally-friendly techniques increasing their use as natural products [49].

Essential oil	Major compounds	Pathogens	References
Mentha longifolia	Menthone (48%); eucalyptol (21.6%)	*Aspergillus flavus*, *Aspergillus niger*, *Fusarium culmorum*	[50]
Salvia dolomitica	1,8-cineole (18.9%) and β-caryophyllene (13.1%)	*Aspergillus niger*, *Aspergillus flavus*, *Trichothecium roseum*	[45]
Mentha arvensis	Menthol (69.2%), Menthone (19.9%)	*Alternaria alternata*	[51]
Rosmarinus officinalis	1,8-Cineole (53.48%) α-Pinene (15.65%)	*Fusarium verticillioides*	[52]
Helianthus annuus	α-Pinene (50.65%)	*Aspergillus niger*, *Candida albicans*, *Cryptococcus neoformans*	[53]
Salvia officinalis	1,8-cineole, α-thujone and camphor, β-caryophyllene, borneol, viridiflorol, α-pinene and camphene	*Botrytis cinerea*, *Penicillium expansum*, *Rhizopus stolonifer*	[54]
Cinnamomum zeylanicum	cinnamaldehyde (52.4%), benzaldehyde (12.31%),	*Aspergillus niger*,*Colletotrichum acutatum*	[55]
Cymbopogon flexousus	Citral (81.84%)	*Colletotrichum acutatum*, *Colletotrichum gloeosporioides*	[56, 57]
Origanum vulgare	Carvacrol (89.98%), β-caryophyllene (3.34%)	*Botrytis cinerea*	[58]

Table 1.
Aromatic plants with potential in management of crop diseases.

Table 1 displays a few EOs with their major compounds and the variety of crops fungal pathogens against which they are effective which were reported from the year 2015. The aromatic plants selected are belonging to families of species that are promising in phytoremediation processes.

5. Disadvantages of essential oils and their formulation

The potential of the use of EOs in organic food production is evident. Although there is considerable evidence of their efficacy, their application is still lacking. Farmers are still reluctant or have not welcomed the use of these products owing to a variety of disadvantages. They are limited due to their high volatility and low stability, low water solubility, strong influence on organoleptic properties, composition variability and phytotoxic effects [58, 59]. The pre-harvest use is further limited because they are very sensitive to light and elevated temperatures will cause oxidation and eventual biodegradation. To overcome the limitations, product formulations are being investigated in pesticides applications. These involve a combination of an active ingredient and inactive materials that act as additives that involve specialized processing of the product to improve its biological qualities, durability, and stability [60]. The formulation for pesticides is performed in two ways, as liquids including aqueous and dispersions and solids including wettable powders and water-dispersible granules or as controlled-release systems [61]. Factors that are considered in formulations include the mode of application, the crop and agricultural practices [62]. The application of nanotechnology for formulation of new products, using polymer-based nanocapsules, or encapsulation with metallic nanoparticles could result in increased stability and efficacy of EOs and therefore reduce the required dosage for application. In encapsulation formulations, EOs are trapped in the carrier matrix in which the release of bioactive components is controlled [63]. Nanoencapsulation is based on encapsulating EOs in materials including lipid nanomaterials, polymeric nanoparticles and clay nanomaterials resulting in improved characteristics [64–66]. The antimicrobial activity of the essential oil-based nanoemulsions is much stronger than in free EOs due to increased surface area which influence the transport of Eos [42]. Recently, a study on the powdered formulations of EOs demonstrated promising results for the control of corn pathogens and evaluating the efficacy of bio-fungicide formulations. Powdered formulations of *Cymbopogon giganteus* and *Eucalyptus camaldulensis* EOs were effective against *Aspergillus flavus*, *Aspergillus parasiticus*, *Aspergillus niger*, *Rhizopus and Fusarium* spp. at concentration ranging between 0.5 and 1.0% dosages highlighting the possibility of EOs in the management of corn diseases [43].

6. Environmental pollution

Mining, industrial and agricultural activities have resulted to significant environmental pollution and land degradation. Overuse of pesticides pollution has resulted to environmental pollution and an increased risk of harmful effects on the ecosystem, and entering into the food chain. The residual concentration of pesticides in agricultural soils is often above the permissible limits by the regulations and therefore the challenge is to reduce these chemicals from the soils [67, 68].

Mining activities generates a big amount of solid mine waste, which contributes enormously to environmental pollution due to the release of potentially toxic elements [69]. In 2022, mine tailings generated worldwide were estimated to be more than 14 billion metric tons, with potential to contaminate soil, water and air [70]. Mine tailings remain in dams/open ponds without further treatment after extraction of valuable minerals and negatively impacting the environment [71]. Abandoned mine tailings damage local land resources and pose severe environmental pollution. Toxic elements including lead, chromium, arsenic, nickel, copper, and cobalt occur in the tailings can enter into the food chain and impact the health of humans and animals [72, 73]. Toxic levels of metal can negatively impact the quality of soil, pollute ground and surface as well as agricultural produce. These poses high risk of food chain contamination and consequently health problems [74]. The influence of heavy metals toxicity effects in living organisms it through their interference in metabolism processes and possibly cause mutagenesis by accumulating in the bodies [75]. They can also be endocrine disruptors, mutagenic, teratogenic while others can cause neurological conditions among children and infants [76].

Generation of electricity has resulted in enormous generation of solid waste which is an important environmental hazard. Solid waste is generated from the combustion of coal in the coal-fired power plants (C-FPPs) and Coal Fly Ash (CFA) is one of the major by-products produced by C-FPPs, which is one of the major global concerns [77]. Its disposal causes significant environmental and economic problems, requiring large quantities of energy, water and land [78]. The disposal of CFA in landfills and ash ponds is the primary management technique, but the fine particles of CFA are dispersed by wind into the atmosphere and remain suspended for a long period of time resulting to air pollution. The fly ash disposal in ash dumps constructed near the power stations could also lead to water and soil contamination through leaching or seepage of pollutants into ground and surface water [79]. CFA disposal sites are hazardous due to the presence of toxic metals such As, Se, Cr, Cd, Pb and Hg and these are continuously being discharged into the environment due to improper disposal of fly ash [80, 81]. These pollution problems can be addressed through phytoremediation strategies that utilize plants species including shrubs, trees and grasses.

7. Phytoremediation: a sustainable approach for addressing environmental pollution

Phytoremediation techniques are cost-effective, environmentally-friendly and effective in the removal of pollutants from pollutes sites through the use of plants [82]. Phytoremediation is the utilization of suitable plant species at polluted areas to remove/reduce the toxic levels of pollutants [83]. Plants reduce the pollutants on the contaminated site by taking up toxicants through their roots and translocating or transforming them into less toxic forms [84]. Phytoremediation is able to remediate polluted areas due to the variety of mechanisms plants species may use to either remove or detoxify contaminant. Strategies include phytodegradation, phytostabilization, phytovolatilization, rhizofiltration, and phytoextraction [85, 86]. The purpose of these approaches differs such as containment, remediation, stabilization, leaching of contaminants, and detoxification [87]. Phytostabilization involves the use of plants that can reduce the movement of pollutants through accumulation by roots [88]. In phytoextraction, the pollutants are transferred to the harvestable plant parts [10]. Phytodegradation involve the breakdown of organic pollutants into less toxic or non-toxic forms through the production of degradation enzymes [89]. The use of plants in restoration processes is a more sustainable and feasible approach as it restores vegetation and prevent erosion, therefore improving the chemical properties of soil overtime.

Ideal species for phytoremediation strategies should have high biomass-producing capabilities, tolerant to toxic effects of metals and contaminants, easy to cultivate, have high absorption capacity. An added advantage includes high value economic crop, with no or low risk of contamination in use of end. Phytoremediation technologies are very appropriate for restoration of soils contaminated with pesticides, mining and industrial activities [90, 91]. Various conventional treatment methods are being used for clean-up of contaminated sites but they are often ineffective, expensive and technically difficult [92]. Chemical and physical techniques also showed to be expensive and not sustainable to the environment. Contaminated sites are remodeled using phytoremediation as a sustainable strategy to lower the pollution load. Plants can help clean up many types of pollutants including metals, radionuclides, and organic pollutants [explosives, nutrients, chlorinated solvents, surfactants, polycyclic aromatic hydrocarbons (PAHs), polychlorinated biphenyls (PCBs), insecticides/pesticides, and various hydrocarbons] [93].

8. Suitability of aromatic plants in phytoremediation process

Recently, aromatic plants have gained interest in phytoremediation strategies and are being explored for their phytoremediation potential and also use of their biomass to extract essential oils that are added-value products. The use of essential oil producing aromatic plants from the Poaceae, Lamiaceae, Asteraceae, and Geraniaceae families can be used for phytoremediation of heavy metal contaminated sites. These plants can act as phytostabilisers, hyper accumulators, bio-monitors, and metallophytes. It has been observed that heavy metal stress enhances the essential oil percentage of certain aromatic crops but that the metals are not present in the essential oils produced [94]. Aromatic grasses received much attention due to their ability to accumulate high biomass and are therefore cultivated for high value essential oil such as Citronella (*Cymbopogon winterianus* Jowitt ex Bor), Lemon grass (*Cymbopogon flexuosus* (Nees ex Steud.) Watson), Vetiver (*Chrysopogon zizanioides* (L.) Nash), and Palmarosa (*Cymbopogon martinii* (Roxb.) Watson) [4, 95]. Grasses provide vegetation cover in a reasonable short period of time and therefore are ideal candidates for restoration of polluted lands. *Cymbopogon winterianus* Jowitt (citronella) and *Cymbopogon flexuosus* (lemon grass) can tolerate soils polluted due to mining activities and organic amendment assist in establishment of these species [96]. *Cymbopogon martinii* can accumulate of toxic concentrations of Cd, Cr, Pb, Ni and can be used for phytoremediation of sewage sludge [97].

Other families including *Ocimum*, *Mentha*, *Lavender*, *Salvia*, *Rosemary* and *Chamomile* have also demonstrated potential in phytoremediation technology. *Lavandula angustifolia* L. has phytoremediation potential for soils polluted with Cu and Pb and the inoculation of arbuscular mycorrhizal fungi results to improved plant growth and essential oil yield [98]. *Ocimum basilicum L.* has phytostabilization potential and can therefore be used to remediate soils polluted with Cd, Co, Cr, Cu, Ni, Pb, and Zn [99]. *Mentha piperita* is a metal tolerant aromatic plant that has the capacity to stabilize As, Cd, Ni, and Pb at the root level demonstrating its phytostabilization potential [100]. *Chamomile* (*Matricaria chamomilla* L) has high tolerance to toxic elements and is a suitable candidate for phytoremediation of soils contaminated with As, Cd, Pb and Zn [101]. Sage (*Salvia* sp. family—Lamiaceae) was found to be an effective hyperaccumulator crop against a variety of heavy metals [102]. *Salvia verbenaca* was evaluated for its potential in the phytorestoration of mine tailings moderately contaminated with copper [103]. In the study, increased accumulation of metals was reported for plants grown in medium treated with compost. Studies to evaluate the potential of scented geraniums, Pelargonium roseum, to uptake and accumulate heavy metals nickel (Ni), cadmium (Cd), or lead (Pb) demonstrated that this species is a hyperaccumulator and is effective in phytoremediation strategies [104]. *Salvia sclarea* L. has potential for remediation of heavily contaminated sites and is classified as Pb hyperaccumulator, and Cd and Zn accumulators [105]. The levels of Pb, Cu and Cd in the essential oil of *Salvia sclarea* L. grown on heavily polluted soils were lower than the accepted maximum permissible concentrations. Cultivation of *Salvia sclarea* in polluted sites increase the abundance of plant-growth promoting rhizobacteria significantly over time, thereby significantly impacting microbial communities in soils [106]. Pot trials evaluated the efficiency of *Lantana camara* in the phytoremediation of Cd, Co, and Pb and the results showed an increased accumulation of metals in the plant parts [107]. Development of vegetation cover on polluted sites is significant because these will prevent air pollution and dispersal of pollutants by wind, reduce mobility and toxicity of heavy metals and improve soil health. The revegetation of ash disposal sites and mine tailings is the best strategy for immobilization of pollutants, thus preventing erosion and leaching of toxic metals into the ecosystem. Aromatic plants have demonstrated potential in remediation of polluted soils and therefore should be explored for other contaminated areas such as coal fly ash

Plant	Mechanism	Elements	Medium	Comments	References
Mentha longifolia	Phytostabilization	Pb, Cd, Cu, Mn, Ni, Zn, Co	Contaminated wetlands	The wild mint showed seasonal fluctuations in accumulation potential with elements being highest in the plant roots	[108]
Salvia sclarea	Phytoextraction	Zn	Contaminated soils	The plant is a Zn accumulator and the tolerance mechanism employed upon exposure to excess Zn include increase in the nutrient uptake, leaf pigment and levels of phenolic compounds	[109]
Thymus daenensis Celak.	Phytoextraction	Pb, Cd	Contaminated soils	The plant is able to absorb significant amounts of Pb and Cd	[110]
Helianthus annus L	Phytoextraction	Pb and Cd	Pb and Cd contaminated soils	The plant was effective in the uptake of Cd and Pb, and can remediate soils polluted with both Pb and Cd	[111]
Alyssum murale L.	Phytoextraction	Ni	Industrial Ni-contaminated soil (heavy clay, sand, organic muck)	Maximum Ni extraction was achieved in *A. murale* grown in unfertilized clay soil accompanied by higher irrigation rate	[112]
Rosmarinus officinalis L.	Phytoextraction	Cd, Zn, Cu, Ni, Cr and Pb	Contaminated soils	The average translocation of metals from soil to root of was found to be in the order of Ni > Pb > Cd > Zn > Cr > Cu	[113]
Matricaria chamomilla L.	Phytoextraction	Cd	Hoagland hydroponic solution	Chamomile is tolerant to Cd stress and is considered a metal excluder	[114]
Vetiver	Phytostabilization	Fe, Mn, Zn, Cu, Pb, Ni, Cr and Al	Contaminated iron ore mine-soil	The plant is an excellent candidate for remediation and restoration of iron-ore mine spoil dumps	[115, 116]

Plant	Mechanism	Elements	Medium	Comments	References
Citronella (Cymbopogon winterianus)	Phytostabilization	Cd	Contaminated soils	The plant can tolerate 50 and 100 mg/kg treatments of cadmium and the growth was inhibited at 200 mg/kg ultimately leading to the plant death	[117]
Chamomile (Matricaria spp.)	Phytoextraction, phytostabilization	As, Pb, Zn, Cd	Contaminated soils	As, Pb, and Zn were retained in roots, Cd showed good ability to translocate to the shoots	[118]
Mentha pipertia (L.)		Zn, Pb, Cd	Contaminated soils	The plant can grow in contaminated soils treated with sewage sludge	[118]

Table 2.
Aromatic plants with phytoremediation potential.

dams that are sparsely reported in literature. Some aromatic species that displayed potential for phytoremediation processes are listed in **Table 2**.

9. Risk assessment of EOs sourced from aromatic plants grown on polluted soils

The use of edible crops for bioremediation strategies is not feasible because the heavy metals can enter into food chain through absorption by crops, and consequently consumption by human or animals. Aromatic crops hold a superior position over food crops for phytoremediation purposes as their use is associated with minimum risk of food chain contamination. Several authors reported that aromatic crops could be grown on heavy metal contaminated sites without causing any significant risks of metal transfer to by-products and alterations in essential oil composition [33, 119]. The essential oil of *Salvia officinalis* cultivated in soils contaminated with metals was found to be free from hazardous heavy metals [120]. These authors reported that heavy metals in EOs extracted from aromatic crops grown on heavy metal contaminated soil were well within the critical limits as specified by FSSA [121]. Essential oils extracted from aromatic plants grown in polluted sites can be used for non-edible products such as soap and detergents manufacturing, cosmetics and perfumery and as insects repellents, therefore minimizing food chain contamination [33]. Hydro-distillation process for essential oil extraction results to less contamination of essential oil by heavy metals [122, 123]. Despite many studies on the use of indigenous species in phytoremediation processes, there remain the knowledge gap with regards to the exploitation of aromatic plants, which present an advantage of producing essential oils that are free from metal toxicity. Aromatic plant resources are very abundant, and they can be used on large scale. These plants offer a novel option for their use in phytoremediation of heavy metal contaminated sites. Extraction of metal-free essential oil will be beneficial to the economy of any nation by exporting these natural products [124].

Essential oil composition is influenced by environmental factors such as; climatic changes, geographical origin, and seasonal factors and consequently influencing the biological properties of the EOs. The composition and biological properties of essential oils of *Helichrsyum splendidum* was influenced by different environmental conditions. Author [125] compared antifungal properties of the EOs sourced from different geographic locations and the bioactivities was influenced by varying levels of major constituents in the respective oils against crops pathogens. Oils characterized with high levels of germacrene d and spathulenol were more active as compared to oils characterized by δ-cadinene and α-cadinol, highlighting that activity is closely linked to the chemical composition of the EOs. Plants grown in polluted soils and fly ash may produce essential oils that may have enhanced production of essential oils and therefore could have improved activity. Many aromatic plants remain unexplored for its potential in phyto-strategies and biological effects. Poplars are tree species that have favorable characteristics for phytoremediation strategies that include quick establishment, fast growth, large biomass accumulation, extensive and deep root systems, high rates of transpiration, ease asexual propagation, grow effortlessly on marginal lands, are not edible and can live for long [126]. Such plants should be investigated more for both their phytoremediation potential and biological properties. *Piper aduncum* has noteworthy activity against *Colletotrichum musae* that causes post-harvest banana fruit rot disease.

This is a fast-growing shrub that thrive in poor soils, dominate degraded forests and abandoned lands [127, 128]. These characteristics make this species ideal for phytoremediation species. As a biopesticide, these can be cultivated in abandoned/contaminated lands followed by extraction of EOs, therefore turning unproductive lands to commercially viable ones.

10. Future prospects

Aromatic plants are important sources of safer and effective agrochemicals and therefore more attention should be given toward their cultivation, preservation, and sustainable development. Mining activities have resulted in significant damages to land and phytoremediation strategies are able to solve problems simultaneous by providing vegetative covers in polluted sites and species to be selected based on their potential for generating revenues. Many aromatic plants are under explored and there is a need to conduct more studies on phytoremediation of these species, with emphasis to those essential oil producing plants. There is an increasing demand of essential oil and aromatic plants can be grown on contaminated sites, especially those that are moderately polluted such as fly ash deposits and as such there is no secondary pollution due to metal toxicity in the essential oils. The cultivation of aromatic crops at heavy metal contaminated sites has often been suggested as a profitable and feasible option. The benefits of using aromatic plants for phytoremediation purpose can be categorized under two main headings as, environmental and economic beneficial. Being high value economic crops, monetary benefits can also be obtained by growing them in contaminated areas.

11. Conclusion

EOs extracted from aromatic plants are more effective for protection and prevention of crops against infectious diseases in the field as well as during storage. EOs are a substantial by-product of aromatic crops that are generally used for non-edible purposes such as the production of soaps and detergents, insect repellents, cosmetics, and scents, and they might be considered a viable option for decreasing food chain contamination. In addition to their potential as biopesticides, aromatic plants have gained a lot of interest in phytoremediation strategies due to their potential for generating income. The use of EOs as biofungicides will help to reduce the chemical fungicides application doses, avoid pathogen resistance and solve problems simultaneously through cultivating of these species in previously unproductive lands. Problem solving strategies should be approached in a manner that support sustainable development of regions, especially in developing countries. Aromatic plants are high value economic crops emerging as candidates for remediation of contaminated sites as they produce essential oils that are free from metal toxicity. Essential oils sourced from plants growing in harsh conditions can lead to production of essential oils with improved activity. Aromatic plants can be cultivated from previously unproductive lands and agricultural soils that are polluted by use of synthetic pesticides at lower cost and therefore reducing costs for farmers. This is an ecologically sustainable method, protecting the environment and restoring the soil structure and providing vegetative cover thereby preventing dispersion of pollutants.

Acknowledgements

The authors would like to thank Tshwane University of Technology for providing the necessary resources.

Conflict of interest

The authors declare no conflict of interest.

Author details

Maria Banda*, Alexis Munyengabe and Wilma Augustyn
Department of Chemistry, Tshwane University of Technology, Pretoria, Gauteng, South Africa

*Address all correspondence to: mashigomf@tut.ac.za

References

[1] Ristaino JR, Anderson PK, Bebber DP, Wei Q, et al. The persistent threat of emerging plant disease pandemics to global food security. Perspective. 2021;**118**:23. DOI: 10.1073/pnas.2022239118

[2] Gajić G, Djurdjević L, Kostić O, Jarić S, Mitrović M, Pavlović P. Ecological potential of plants for phytoremediation and ecorestoration of fly ash. Environmental Science and Technology. 2018;**6**:124. DOI: 10.3389/fenvs.2018.00124

[3] Maiti D, Prasad B. Revegetation of fly ash – A review with emphasis on grass-legume plantation and bioaccumulation of metals. Applied Ecology and Environmental Research. 2016;**2016**:185-212. DOI: 10.15666/aeer/1402_185212

[4] Mishra B, Chandra M. Evaluation of phytoremediation potential of aromatic plants: A systematic review. Journal of Applied Research on Medicinal and Aromatic Plants. 2022;**31**:100405. DOI: 10.1016/j.jarmap.2022.100405

[5] Nnadi NE, Carter DA. Climate change and the emergence of fungal pathogens. PLoS Pathogens. 2021;**17**(4):1-6. DOI: 10.1371/journal.ppat.1009503

[6] Arias-Estévez M, López-Periago E, Martínez-Carballo E, Simal-Gándara J, Mejuto JC, García-Río L. The mobility and degradation of pesticides in soils and the pollution of groundwater resources. Agriculture, Ecosystems & Environment. 2008;**123**(4):247-260. DOI: 10.1016/j.agee.2007.07.011

[7] Olutona GO, Aderemi MA. Organochlorine pesticide residue and heavy metals in leguminous food crops from selected markets in Ibadan, Nigeria. Legume. Science. 2019;**1**(1). DOI: 10.1002/leg3.3

[8] Jayaraj R, Megha P, Sreedev P. Organochlorine pesticides, their toxic effects on living organisms and their fate in the environment. Interdisciplinary Toxicology. 2016;**9**(3-4):90-100. DOI: 10.1515/intox-2016-0012

[9] UNEP (United Nations Environmental Programme). Stockholm Convention on Persistent Organic Pollutants (POPs), Secretariat of the Stockholm Convention. 2013. Available from: http://chm.pops.int/portals/0/repository/convention_text/unep-pops-cop-convtext-full.english.pdf [Accessed: April 21, 2023]

[10] Kumar B, Verma VK, Mishra M, Kumar S, Sharma CS, Akolkar AB. Persistent organic pollutants in residential soils of North India and assessment of human health hazard and risks. Toxicological & Environmental Chemistry. 2014;**96**(2):255-272. DOI: 10.1080/02772248.2014.923427

[11] Gunstone T, Cornelisse T, Klein K, Dubey A, Donley N. Pesticides and soil invertebrates: A hazard assessment. Frontiers in Environmental Science. 2021;**122**:1-21. DOI: 10.3389/fenvs.2021.643847

[12] Bauer I, Misslinger M, Shadkchan Y, Dietl AM, Petzer V, Orasch T, et al. The lysine deacetylase RpdA is essential for virulence in *Aspergillus fumigatus*. Frontiers in Microbiology. 2019;**10**:2773. DOI: 10.3389/fmicb.2019.02773

[13] Aikpokpodion PE, Lajide L, Aiyesanmi AF. Impacts of Cu-based fungicide on copper residue and mineral elements distribution in cocoa beans and pods. World Journal of Agricultural

Sciences. 2013;**9**(1):10-16. DOI: 10.5829/idosi.wjas.2013.9.1.1133

[14] Lamichhane JR, Osdaghi E, Behlau F, Köhl J, Jones JB, Aubertot JN. Thirteen decades of antimicrobial copper compounds applied in agriculture. A review. Agronomy for Sustainable Development. 2018;**38**(3):1-18. DOI: 10.1007/s13593-018-0503-9

[15] Grzanka M, Sobiech Ł, Danielewicz J, Horoszkiewicz-Janka J, Skrzypczak G, Sawinska Z, et al. Impact of essential oils on the development of pathogens of the Fusarium genus and germination parameters of selected crops. Open Chemistry. 2021;**19**(1):884-893. DOI: 10.1515/chem-2021-0079

[16] Kesraoui S, Andrés MF, Berrocal-Lobo M, Soudani S, Gonzalez-Coloma A. Direct and indirect effects of essential oils for sustainable crop protection. Plants. 2022;**11**(16):2144. DOI: 10.3390/plants11162144

[17] Seow YX, Yeo CR, Chung HL, Yuk HG. Plant essential oils as active antimicrobial agents. Critical Reviews in Food Science and Nutrition. 2014;**54**(5):625-644. DOI: 10.1080/10408398.2011.599504

[18] Reddy KRN, Salleh B, Saad B, Abbas HK, Abel CA, Shier WT. An overview of mycotoxin contamination in foods and its implications for human health. Toxin Reviews. 2010;**29**(1):3-26. DOI: 10.3109/15569541003598553

[19] Chang Y, Harmon PF, Treadwell DD, Carrillo D, Sarkhosh A, Brecht JK. Biocontrol potential of essential oils in organic horticulture systems: From farm to fork. Frontiers in Nutrition. 2022;**8**:1275. DOI: 10.3389/fnut.2021.805138

[20] Verweij PE, Chowdhary A, Melchers WJ, Meis JF. (2016). Azole resistance in aspergillus fumigatus: Can we retain the clinical use of mold-active antifungal azoles? Clinical Infectious Diseases. 2016;**62**(3):362-368. DOI: 10.1093/cid/civ885

[21] Loi M, Paciolla C, Logrieco AF, Mulè G. Plant bioactive compounds in pre-and postharvest management for aflatoxins reduction. Frontiers in Microbiology. 2020;**11**:243. DOI: 10.3389/fmicb.2020.00243

[22] Berger S, El Chazli Y, Babu AF, Coste AT. Azole resistance in *Aspergillus fumigatus*: A consequence of antifungal use in agriculture? Front. Microbiology. 2017;**7**(8):1024. DOI: 10.3389/fmicb.2017.01024

[23] Bosetti D, Neofytos D. Invasive Aspergillosis and the impact of azole-resistance. Current Fungal Infection Reports. 2023;**2023**:1-10. DOI: 10.1007/s12281-023-00459-z

[24] Klich MA. Aspergillus flavus: The major producer of aflatoxin. Molecular Plant Pathology. 2007;**8**(6):713-722. DOI: 10.1111/j.1364-3703.2007.00436.x

[25] Langfeldt A, Gold JA, Chiller T. Emerging fungal infections: From the fields to the clinic, resistant aspergillus fumigatus and dermatophyte species: A one health perspective on an urgent public health problem. Current Clinical Microbiology Reports. 2022;**9**(4):46-51. DOI: 10.1007/s40588-022-00181-3

[26] Saremi H, Okhovvat SM. Major Fusarium diseases on crops and their control management with soil solarisation in Northwest Iran. Communications in Agricultural and Applied Biological Sciences. 2008;**73**(2):189-199

[27] Arie T. Fusarium diseases of cultivated plants, control, diagnosis, and molecular and genetic studies. Journal of Pesticide Science. 2019;**25**:275-281. DOI: 10.1584/jpestics

[28] Sáenz V, Alvarez-Moreno C, Pape PL, Restrepo S, Guarro J, Ramírez AMC. A one health perspective to recognize Fusarium as important in clinical practice. Journal of Fungi. 2020;**6**(4):235. DOI: 10.3390/jof6040235

[29] Zhao B, He D, Wang L. Advances in Fusarium drug resistance research. Journal of Global Antimicrobial Resistance. 2021;**24**:215-219. DOI: 10.1016/j.jgar.2020.12.016

[30] Al-Hatmi AM, de Hoog GS, Meis JF. Multiresistant Fusarium pathogens on plants and humans: Solutions in (from) the antifungal pipeline? Infect Drug Resistances. 2019;**28**(12):3727-3737. DOI: 10.2147/IDR.S180912

[31] Rupp S, Plesken C, Rumsey S, Dowling M, Schnabel G, Weber RW, et al. Botrytis fragariae, a new species causing gray mold on strawberries, shows high frequencies of specific and efflux-based fungicide resistance. Applied and Environmental Microbiology. 2017;**83**(9):1-16. DOI: 10.1128/AEM.00269-17

[32] Shao W, Zhao Y, Ma Z. Advances in understanding fungicide resistance in Botrytis cinerea in China. Phytopathology. 2021;**111**(3):455-463. DOI: 10.1094/PHYTO-07-20-0313-IA

[33] Lal K, Yadav RK, Kaur R, Bundela DS, Khan MI, Chaudhary M, et al. Productivity, essential oil yield, and heavy metal accumulation in lemon grass (*Cymbopogon flexuosus*) under varied wastewater-groundwater irrigation regimes. Industrial Crops and Products. 2013;**45**:270-278. DOI: 10.1016/j.indcrop.2013.01.004

[34] Schepers HT, Kessel GJ, Lucca F, Förch MG, Van Den Bosch GBM, Topper CG, et al. Reduced efficacy of fluazinam against Phytophthora infestans in the Netherlands. European Journal of Plant Pathology. 2018;**151**:947-960. DOI: 10.1007/s10658-018-1430-y

[35] Lichtner FJ, Gaskins VL, Cox KD. Global transcriptomic responses orchestrate difenoconazole resistance in Penicillium spp. causing blue mold of stored apple fruit. BMC Genomics. 2020;**21**:574. DOI: 10.1186/s12864-020-06987-z

[36] Bhatta UK. Alternative management approaches of citrus diseases caused by *Penicillium digitatum* (green mold) and *Penicillium italicum* (blue mold). Frontiers in Plant Science. 2022;**12**:833328. DOI: 10.3389/fpls.2021.833328

[37] Palou L. *Penicillium digitatum*, *Penicillium italicum* (green mold, blue mold). In: Postharvest Decay. London: Academic Press; 2014. pp. 45-102. DOI: 10.1016/B978-0-12-411552-1.00002-8

[38] Wallace RL, Hirkala DL, Nelson LM. Postharvest biological control of blue mold of apple by *Pseudomonas fluorescens* during commercial storage and potential modes of action. Postharvest Biology and Technology. 2017;**133**:1-11. DOI: 10.1016/j.postharvbio.2017.07.003

[39] Erasmus A, Lennox CL, Korsten L, Lesar K, Fourie PH. Imazalil resistance in *Penicillium digitatum* and *P. italicum* causing citrus postharvest green and blue mould: Impact and options. Postharvest Biology and Technology. 2015;**107**:66-76. DOI: 10.1016/j.postharvbio.2015.05.008

[40] Papoutsis K, Mathioudakis M, Hasperué JH, Ziogas V. Non-chemical treatments for preventing the

postharvest fungal rotting of citrus caused by *Penicillium digitatum* (green mold) and *Penicillium italicum* (blue mold). Trends in Food Science & Technology. 2019;**86**:479-491. DOI: 10.1016/j.tifs.2019.02.053

[41] Miastkowska M, Michalczyk A, Figacz K, Sikora E. Nanoformulations as a modern form of biofungicide. Journal of Environmental Health Science and Engineering. 2020;**18**:119-128. DOI: 10.1007/s40201-020-00445-4

[42] Ouattara LP, Dindane Z, Sawadogo I, Soala WR, Zida PE, Konate K, et al. Antifungal activity of essential oil-based formulations used in corn preservation in Burkina Faso. African Journal of Microbiology Research. 2022;**16**(10):327-333. DOI: 10.5897/AJMR2022.9662

[43] Ebani VV, Nardoni S, Bertelloni F, Giovanelli S, Ruffoni B, D'Ascenzi C, et al. Activity of *Salvia dolomitica* and *Salvia somalensis* essential oils against bacteria, molds and yeasts. Molecules. 2018;**23**(2):396. DOI: 10.3390/molecules23020396

[44] Ben Abdallah S, Riahi C, Vacas S, Navarro-Llopis V, Urbaneja A, Hedo P. The dual benefit of plant essential oils against Tuta absoluta. Plants. 2020;**12**(5):985. DOI: 10.3390/plants12050985

[45] Singh P, Pandey AK. A review: Prospective of essential oils of the genus Mentha as biopesticides. Frontiers in Plant Science. 2018;**9**(1295):1-14. DOI: 10.3389/fpls.2018.01295

[46] Perczak A, Gwiazdowska D, Marchwińska K, Juś K, Gwiazdowski R, Waśkiewicz A. Antifungal activity of selected essential oils against Fusarium culmorum and Fusarium graminearum and their secondary metabolites in wheat seeds. Archives of Microbiology. 2019;**201**(8):1085-1097. DOI: 10.1007/s00203-019-01673-5

[47] Chouhan S, Sharma K, Guleria S. Antimicrobial activity of some essential oils—Present status and future perspectives. Medicine. 2017;**4**(3):58. DOI: 10.3390/medicines4030058

[48] Fierascu RC, Fierascu I, Avramescu SM, Sieniawska E. Recovery of natural antioxidants from agro-industrial side streams through advanced extraction techniques. Molecules. 2019;**24**(23):4212. DOI: 10.3390/molecules24234212

[49] Carrubba A, Catalano C. Essential oil crops for sustainable agriculture–a review. Climate Change, Intercropping, Pest Control and Beneficial Microorganisms. 2009;**2009**:137-187. DOI: 10.1007/978-90-481-2716-08

[50] Ali HM, Elgat WAAA, El-Hefny M, Salem MZM, Taha AS, Al Farraj DA, et al. New approach for using of *Mentha longifolia* L. and *Citrus reticulata* L. essential oils as wood-biofungicides: GC-MS, SEM, and MNDO quantum chemical studies. Materials (Basel). 2021;**14**(6):1361. DOI: 10.3390/ma14061361

[51] Perveen K, Bokahri NA. Management of Alternaria leaf blight in tomato plants by mentha essential oil. Plant Protection Science. 2020;**56**(3):191-196. DOI: 10.17221/100/2019-PPS

[52] Chimón F, Vanessa B, Pizzolitto R, Sanchez A, Gómez E, Zygadlo J. Chemical composition and antifungal properties of commercial essential oils against the maize phytopathogenic fungus *Fusarium verticillioides*. Revista Argentina de Microbiología. 2021;**2021**:53. DOI: 10.1016/j.ram.2020.12.001

[53] Lawson SK, Sharp LG, Powers CN, McFeeters RL, Satyal P, Setzer WN.

Essential oil compositions and antifungal activity of sunflower (Helianthus) species growing in North Alabama. Applied Sciences. 2019;**9**(15):3179. DOI: 10.3390/app9153179

[54] Wei ZF, Zhao RN, Dong LJ, Zhao XY, Su JX, Zhao M, et al. Dual-cooled solvent-free microwave extraction of *Salvia officinalis* L. essential oil and evaluation of its antimicrobial activity. Industrial Crops and Products. 2018;**120**:71-76. DOI: 10.1016/j.indcrop.2018.04.058

[55] He J, Wu D, Zhang Q, Chen H, Li H, Han Q, et al. Efficacy and mechanism of cinnamon essential oil on inhibition of Colletotrichum acutatum isolated from 'Hongyang'kiwifruit. Frontiers in Microbiology. 2018;**2018**(9):1288. DOI: 10.3389/fmicb.2018.01288

[56] Mukarram M, Khan M, Corpas FJ. Silicon nanoparticles elicit an increase in lemongrass (*Cymbopogon flexuosus* (Steud.) Wats) agronomic parameters with a higher essential oil yield. Journal of Hazardous Materials. 2021;**412**:125254. DOI: 10.1016/j.jhazmat.2021.125254

[57] Antonioli G, Fontanella G, EcheverrigarayS,DelamareAPL,PaulettiGF, BarcellosT. Poly (lactic acid) nanocapsules containing lemongrass essential oil for postharvest decay control: In vitro and in vivo evaluation against phytopathogenic fungi. Food Chemistry. 2020;**326**:126997. DOI: 10.1016/j.foodchem.2020.126997

[58] Hou H, Zhang X, Zhao T, Zhou L. Effects of Origanum vulgare essential oil and its two main components, carvacrol and thymol, on the plant pathogen Botrytis cinerea. PeerJ. 2020;**8**:e9626. DOI: 10.7717/peerj.9626

[59] Chang Y, McLandsborough L, McClements DJ. Fabrication, stability and efficacy of dual-component antimicrobial nanoemulsions: Essential oil (thyme oil) and cationic surfactant (lauric arginate). Food Chemistry. 2015;**172**:298-304. DOI: 10.1016/j.foodchem.2014.09.081

[60] Libs E, Salim E. Formulation of essential oil pesticides technology and their application. Agricultural Research & Technology. 2017;**9**(2):555759. DOI: 10.19080/ARTOAJ.2017.09.555759

[61] Markus A. Advances in the technology of controlled-release pesticide formulations. Drugs and the Pharmaceutical Sciences. 1996;**73**:73-91

[62] Moretti M, Marcarelli M, Villarini M, Fatigoni C, Scassellati-Sforzolini G, Pasquini R. In vitro testing for genotoxicity of the herbicide terbutryn: Cytogenetic and primary DNA damage. Toxicology in Vitro. 2002;**16**(1):81-88. DOI: 10.1016/S0887-2333(01)00092-3

[63] Liao W, Badri W, Dumas E, Ghnimi S, Elaissari A, Saurel R, et al. Nanoencapsulation of essential oils as natural food antimicrobial agents: An overview. Applied Sciences. 2021;**11**:5778. DOI: 10.3390/app11135778

[64] Chaudhari AK, Singh VK, Kedia A, Das S, Dubey NK. Essential oils and their bioactive compounds as eco-friendly novel green pesticides for management of storage insect pests: Prospects and retrospects. Environmental Science and Pollution Research. 2021;**28**:18918-18940. DOI: 10.1007/S11356-021-12841-W

[65] Esmaili F, Sanei-Dehkordi A, Amoozegar F, Osanloo M. A review on the use of essential oil-based nanoformulations in control of mosquitoes. Biointerface Research in Applied Chemistry. 2021;**11**(5):12516-12529. DOI: 10.33263/BRIAC115.1251612529

[66] Garrido-Miranda KA, Giraldo JD, Schoebitz M. Essential oils and

their formulations for the control of Curculionidae pests. Frontiers in Agronomy. 2022;**4**:876687. DOI: 10.3389/fagro.2022.876687

[67] Silva V, Mol HG, Zomer P, Tienstra M, Ritsema CJ, Geissen V. Pesticide residues in European agricultural soils–a hidden reality unfolded. Science of the Total Environment. 2019;**653**:1532-1545. DOI: 10.1016/j.scitotenv.2018.10.441

[68] Sandanayake S, Hettithanthri O, Buddhinie PKC, Vithanage M. Plant uptake of pesticide residues from agricultural soils. In: Pesticides in Soils: Occurrence, Fate, Control and Remediation. Cham: Springer International Publishing; 2022. p. 197. DOI: 10.1007/698_2021_806

[69] Azizi M, Faz A, Zornoza R, Martinez-Martinez S, Acosta JA. Phytoremediation potential of native plant species in mine soils polluted by metal (loid) s and rare earth elements. Plants. 2023;**12**(6):1219. DOI: 10.3390/plants12061219

[70] Cacciuttolo C, Cano D. Environmental impact assessment of mine tailings spill considering metallurgical processes of Gold and copper mining: Case studies in the Andean countries of Chile and Peru. Water. 2022;**14**(19):3057. DOI: 10.3390/w14193057

[71] Sarathchandra SS, Rengel Z, Solaiman ZM. A review on remediation of Iron ore mine tailings via organic amendments coupled with phytoremediation. Plants. 2023;**12**(9):1871. DOI: 10.3390/plants12091871

[72] Wang L, Ji B, Hu Y, Liu R, Sun W. A review on in situ phytoremediation of mine tailings. Chemosphere. 2017;**184**:594-600. DOI: 10.1016/j.chemosphere.2017.06.025

[73] Shi J, Qian W, Jin Z, Zhou Z, Wang X, Yang X. Evaluation of soil heavy metals pollution and the phytoremediation potential of copper-nickel mine tailings ponds. PLoS One. 2023;**18**(3):e0277159. DOI: 10.1371/journal.pone.0277159

[74] Blaylock MJ. Phytoextraction of metals. In: Phytoremediation of Toxic Metals: Using Plants to Clean up the Environment. New York: Wiley; 2000. pp. 53-70

[75] Pandey G, Madhuri S. Heavy metals causing toxicity in animals and fishes. Research Journal of Animal, Veterinary and Fishery Sciences. 2014;**2**(2):17-23. Available from: http://www.isca.in, www.isca.me [Accessed: April 24, 2023]

[76] Mahar A, Wang P, Ali A, Awasthi MK, Lahori AH, Wang Q, et al. Challenges and opportunities in the phytoremediation of heavy metals contaminated soils: A review. Ecotoxicology and Environmental Safety. 2016;**126**:111-121. DOI: 10.1016/j.ecoenv.2015.12.023

[77] Yadav VK, Gacem A, Choudhary N, Rai A, Kumar P, Yadav KK, et al. Status of coal-based thermal power plants, coal Fly ash production, utilization in India and their emerging applications. Minerals. 2022;**12**(12):1503. DOI: 10.3390/min12121503

[78] Ghosh M, Paul J, Jana A, De A, Mukherjee A. Use of the grass, *Vetiveria zizanioides* (L.) Nash for detoxification and phytoremediation of soils contaminated with fly ash from thermal power plants. Ecological Engineering. 2015;**74**:258-265. DOI: 10.1016/j.ecoleng.2014.10.011

[79] Gajić G, Djurdjević L, Kostić O, Jarić S, Mitrović M, Pavlović P. Ecological potential of plants for phytoremediation and ecorestoration of fly ash deposits and mine wastes. Frontiers in Environmental Science. 2018;**6**:124. DOI: 10.3389/fenvs.2018.00124

[80] Ram LC, Masto RE, Srivastava NK, George J, Selvi VA, Das TB, et al. Potentially toxic elements in lignite and its combustion residues from a power plant. Environmental Monitoring and Assessment. 2015;**187**:1-14. DOI: 10.1007/s10661-014-4148-0

[81] Kostić O, Gajić G, Jarić S, Vukov T, Matić M, Mitrović M, et al. An assessment of the phytoremediation potential of planted and spontaneously colonized woody plant species on chronosequence fly ash disposal sites in Serbia-case study. Plants. 2022;**11**(1):110. DOI: 10.3390/plants11010110

[82] Prasad K, Kumar H, Singh L, Sawarkar AD, Kumar M, Kumar S. Phytocapping technology for sustainable management of contaminated sites: Case studies, challenges, and future prospects. In: Phytoremediation Technology for the Removal of Heavy Metals and Other Contaminants from Soil and Water. Amsterdam, Netherlands: Elsevier; 2022. p. 601. DOI: 10.1016/B978-0-323-85763-5.00041-6

[83] Landmeyer JE. Introduction to Phytoremediation of Contaminated Groundwater: Historical Foundation, Hydrologic Control, and Contaminant Remediation. London, New York: Springer Science & Business Media; 2011. p. 1. DOI: 10.1007/978-3-642-35564-6_1

[84] Meagher RB. Phytoremediation of toxic elemental and organic pollutants. Current Opinion in Plant Biology. 2000;**2**:153-162. DOI: 10.1016/S1369-5266(99)00054-0

[85] Lee JH. An overview of phytoremediation as a potentially promising technology for environmental pollution control. Biotechnology and Bioprocess Engineering. 2013;**18**:431-439. DOI: 10.1007/s12257-013-0193-8

[86] Nedjimi B. Phytoremediation: A sustainable environmental technology for heavy metals decontamination. SN Applied Sciences. 2021, 2021;**3**(3):286. DOI: 10.1007/s42452-021-04301-4

[87] Kamusoko R, Jingura RM. Utility of Jatropha for phytoremediation of heavy metals and emerging contaminants of water resources: A review. Clean-Soil, Air, Water. 2017;**45**(11):1700444. DOI: 10.1002/clen.201700444

[88] Bolan NS, Park JH, Robinson B, Naidu R, Huh KY. Phytostabilization: A green approach to contaminant containment. Advances in Agronomy. 2011;**112**:145-204. DOI: 10.1016/B978-0-12-385538-1.00004-4

[89] Kafle A, Timilsina A, Gautam A, Adhikari K, Bhattarai A, Aryal N. Phytoremediation: Mechanisms, plant selection and enhancement by natural and synthetic agents. Environmental Advances. 2022;**2022**:100203. DOI: 10.1016/j.envadv.2022.100203

[90] Ramírez-Zamora J, Mussali-Galante P, Rodríguez A, Castrejón-Godínez ML, Valencia-Cuevas L, Tovar-Sánchez E. Assisted-Phytostabilization of mine-tailings with Prosopis laevigata (Fabaceae) and biochar. Plants. 2022;**11**(24):3441. DOI: 10.3390/plants11243441

[91] Tarla DN, Erickson LE, Hettiarachchi GM, Amadi SI, Galkaduwa M, Davis L, et al. Phytoremediation and bioremediation of pesticide-contaminated soil. Applied Sciences. 2020;**10**(4):1217. DOI: 10.3390/app10041217

[92] Sharma S, Tiwari S, Hasan A, Saxena V, Pandey LM. Recent advances in conventional and contemporary methods for remediation of heavy metal-contaminated soils. Biotechnology.

2018;**8**(4):216. DOI: 10.1007/s13205-018-1237-8

[93] Raskin I, Ensley BD. Phytoremediation of Toxic Metals. New York: John Wiley & Sons, Inc.; 2000. p. 53

[94] Pandey J, Verma RK, Singh S. Suitability of aromatic plants for phytoremediation of heavy metal contaminated areas: A review. International Journal of Phytoremediation. 2019;**21**(5):405-418. DOI: 10.1080/15226514.2018.1540546

[95] Maddhesiya PK, Singh K, Singh RP. Effects of perennial aromatic grass species richness and microbial consortium on soil properties of marginal lands and on biomass production. Land Degradation & Development. 2021;**32**(2):1008-1021. DOI: 10.1002/ldr.3742

[96] Dowarah J, Boruah D, Gogoi J, Pathak N, Saikia N, Handique AK. Eco-restoration of a high-sulphur coal mine overburden dumping site in Northeast India: A case study. Journal of Earth System Science. 2009;**118**:597-608. DOI: 10.1007/s12040-009-0042-5

[97] Singh G, Pankaj U, Ajayakumar PV, Verma RK. Phytoremediation of sewage sludge by *Cymbopogon martinii* (Roxb.) Wats. var. motia Burk. grown under soil amended with varying levels of sewage sludge. International Journal of Phytoremediation. 2020;**22**(5):540-550. DOI: 10.1080/15226514.2019.1687422

[98] PirsarandibY,HassanpouraghdamMB, Rasouli F, Aazami MA, Puglisi I, Baglieri A. Phytoremediation of soil contaminated with heavy metals via arbuscular mycorrhiza (Funneliformis mosseae) inoculation ameliorates the growth responses and essential oil content in lavender (*Lavandula angustifolia* L.). Agronomy. 2022;**12**(5):1221. DOI: 10.3390/agronomy12051221

[99] Dinu C, Vasile GG, Buleandra M. Translocation and accumulation of heavy metals in *Ocimum basilicum* L. plants grown in a mining-contaminated soil. Journal of Soils Sediments. 2020;**20**:2141-2154. DOI: 10.1007/s11368-019-02550-w

[100] Dinu C, Gheorghe S, Tenea AG, Stoica C, Vasile GG, Popescu RL, et al. Toxic metals (As, Cd, Ni, Pb) impact in the most common medicinal plant (*Mentha piperita*). International Journal of Environmental Research and Public Health. 2021;**18**(8):3904. DOI: 10.3390/ijerph18083904

[101] Száková J, Dziaková M, Kozáková A, Tlustoš P. The risk element uptake by chamomile (*Matricaria recutita* (L.) Rauschert) growing in four different soils. Archives of Environmental Protection. 2018;**44**(4). DOI: 10.24425/122298

[102] Chandra R, Saxena G, Kumar V. Phytoremediation of environmental pollutants: An eco-sustainable green technology to environmental management. In: Advances in Biodegradation and Bioremediation of Industrial W. Florida, FL, United States of America: CRC Press; 2015. p. 12

[103] Novo L, Covelo E, González L. The potential of *Salvia verbenaca* for phytoremediation of copper mine tailings amended with technosol and compost. Water Air and Soil Pollution. 2013;**224**:1513. DOI: 10.1007/s11270-013-1513-5

[104] Mahdieh M, Yazdani M, Mahdieh S. The high potential of Pelargonium roseum plant for phytoremediation of heavy metals. Environmental Monitoring and

Assessment. 2013;**185**:7877-7881. DOI: 10.1007/s10661-013-3141-3

[105] Angelova VR, Ivanova RV, Todorov GM, Ivanov KI. Potential of *Salvia sclarea* L. for phytoremediation of soils contaminated with heavy metals. International Journal of Agricultural and Biosystems Engineering. 2016;**10**(12):780-790. DOI: 10.5281/zenodo.1127523

[106] Raveau R, Fontaine J, Hijri M, Lounes-Hadj SA. The aromatic plant clary sage shaped bacterial communities in the roots and in the trace element-contaminated soil more than mycorrhizal inoculation–a two-year monitoring field trial. Frontiers in Microbiology. 2020;**11**:586050. DOI: 10.3389/fmicb.2020.586050

[107] Ibrahim M. In vitro accumulation potentials of heavy metals in big-sage (*Lantana camara* L.) plant. Dysona Life Science. 2021;**2**:12-18. DOI: 10.30493/dls.2021.254481

[108] Gharib FA, Mansour KH, Ahmed EZ, Galal TM. Heavy metals concentration, and antioxidant activity of the essential oil of the wild mint (*Mentha longifolia* L.) in the Egyptian watercourses. International Journal of Phytoremediation. 2021;**23**(6):641-651. DOI: 10.1080/15226514.2020.1847035

[109] Dobrikova A, Apostolova E, Hanć A, Yotsova E, Borisova P, Sperdouli I, et al. Tolerance mechanisms of the aromatic and medicinal plant *Salvia sclarea* L. to excess zinc. Plants (Basel). 2021;**10**(2):194. DOI: 10.3390/plants10020194

[110] Mohkami Z, Bidarnamani F, Forouzandeh M, Ghafari MZ. The effect of AFM Fungi on Lead and cadmium phytoremediation by thyme (*Thymus daenensis Celak.*). Journal of Water and Soil Conservation. 2018;**25**(4):225-242. DOI: 10.22069/jwsc.2018.12257.2677

[111] Alaboudi KA, Ahmed B, Brodie G. Phytoremediation of Pb and Cd contaminated soils by using sunflower (*Helianthus annuus*) plant. Annals of Agricultural Sciences. 2018;**63**(1):123-127. DOI: 10.1016/j.aoas.2018.05.007

[112] Dehghani S, Zupfer KR, Vasiluk L, Dutton MD, Bellantino-Perco M, Hale BA. Modeling phytoremediation of aged soil Ni from anthropogenic deposition using *Alyssum murale*. Chemosphere. 2021;**267**:128861. DOI: 10.1016/j.chemosphere.2020.128861

[113] Ziarati P, Iranzad-Asl S, Asgarpanah J. Companion Pelargonium roseum and Rosmarinus officinalis in cleaning up contaminated soil by phytoextraction technique the role of companion plants in boosting phytoremediation potential. International Journal of Plant, Animal and Environmental Sciences. 2014;**4**(3):424-430. Available from: www.ijpaes.com

[114] Kováčik J, Tomko J, Bačkor M, Repčák M. Matricaria chamomilla is not a hyperaccumulator, but tolerant to cadmium stress. Plant Growth Regulation. 2006;**50**:239-247. DOI: 10.1007/s10725-006-9141-3

[115] Banerjee R, Goswami P, Pathak K, Mukherjee A. Vetiver grass: An environment clean-up tool for heavy metal contaminated iron ore mine-soil. Ecological Engineering. 2016;**90**:25-34. DOI: 10.1016/j.ecoleng.2016.01.027

[116] Suelee AL, Hasan SN, Kusin FM, Yusuff FM, Ibrahim ZZ. Phytoremediation potential of vetiver grass (*Vetiveria zizanioides*) for treatment of metal-contaminated water. Water, Air, & Soil Pollution. 2017;**228**:1-15. DOI: 10.1007/s11270-017-3349-x

[117] Khilji S, Sajid Z. Phytoremediation potential of lemongrass (*Cymbopogon flexuosus* stapf.) Grown on tannery sludge contaminated soil. Applied Ecology and Environmental Research. 2020;**18**:7703-7715. DOI: 10.15666/aeer/1806_77037715

[118] Mohseni A, Reyhanitabar A, Nosratollah N, Shahin O, Kambiz B. Phytoremediation potential and essential oil quality of peppermint grown in contaminated soils as affected by sludge and time. Journal of Agricultural Science and Technology. 2022;**24**:709-723. DOI: https://jast.modares.ac.ir/article-23-43328-en.html

[119] Zheljazkov VD, Craker LE, Xing B. Effects of Cd, Pb and Cu on growth and essential oil contents in dill, peppermint, and basil. Environmental and Experimental Botany. 2006;**58**(1-3):9-16. DOI: 10.1016/j.envexpbot.2005.06.008

[120] Abu-Darwish MS, Al-Fraihat AH, Al-Dalain SY, Afifi F, Al-Tabbal JA. Determination of essential oils and heavy metals accumulation in *Salvia officinalis* cultivated in three intra-raw spacing in Ash-Shoubak, Jordan. International Journal of Agriculture & Biology. 2011;**13**(6):981-985

[121] Fssai I. Food safety and standards (contaminants, toxins and residues) regulations. Ministry of Health and Family Welfare, India. 2011;**2011**:2. Available from: https://www.fssai.gov.in/upload/uploadfiles/files/Compendium_Contaminants_Regulations_20_08_2020.pdf [Accessed: April 12, 2023]

[122] Scora RW, Chang AC. Essential oil quality and heavy metal concentrations of peppermint grown on a municipal sludge-amended soil. American Society of Agronomy Crop Science Society of America, and Soil Science Society of America. 1997;**26**(4):975-979. DOI: 10.2134/jeq1997.00472425002600040007x

[123] Bernstein N, Chaimovitch D, Dudai N. Effect of irrigation with secondary treated effluent on essential oil, antioxidant activity, and phenolic compounds in oregano and rosemary. Agronomy Journal. 2009;**101**(1):1-10. DOI: 10.2134/agronj2007.0144

[124] Gupta AK, Verma SK, Khan K, Verma RK. Phytoremediation using aromatic plants: A sustainable approach for remediation of heavy metals polluted sites. 2013. 10.1021/es403469c

[125] Mashigo M, Combrinck S, Regnier T, Du Plooy W, Augustyn W, Mokgalaka N. Chemical variations, trichome structure and antifungal activities of essential oils of Helichrysum splendidum from South Africa. South African Journal of Botany. 2015;**96**:78-84. DOI: 10.1016/j.sajb.2014.10.006

[126] Guerra F, Gainza F, Pérez R, Zamudio F. Phytoremediation of heavy metals using poplars (Populus spp.): a glimpse of the plant responses to copper, cadmium and zinc stress. In: Handbook of Phytoremediation. New York: Nova Science; 2011. p. 387

[127] Idris H, Suryani E, Gustia H, Ramadhan AI. The effect of various essential oil and solvent additives on the botanical pesticide of Piper Aduncum essential oil on formulation antifungal activity. Results in Engineering. 2022;**16**:100644. DOI: 10.1016/j.rineng.2022.100644

[128] Cunningham SD, Ow DW. Promises and prospects of phytoremediation. Plant Physiology. 1996;**110**(3):715. DOI: 10.1104%2Fpp.110

Chapter 9

A Review on Vegetable Oil Refining: Process, Advances and Value Addition to Refining by-Products

Anup Sonawane and Samadhan R. Waghmode

Abstract

Nowadays, consumer food choices are driven by health awareness and sustainability concerns. As vegetable oil is an important component of the human diet, the source and the processing play an important role in consumer acceptability. To remove impurities that affect the color, palatability, stability, and safety of oil, crude vegetable oil must be refined. This review highlights the processes and steps used in vegetable oil refining. Depending upon the oil source type, either chemical or physical refining is employed to get the desired oil specifications. Oil refining steps are sequential, with each step removing one or more specific impurities. Refining advances aim towards minimizing chemical usage, nutrient losses, oil losses, and avoiding the formation of *trans*-fatty acids. The review also discusses the prospect of using the refining by-product stream for obtaining high-value products like phosphatidylcholine, tocopherols, and tocotrienols. The edible oil industry can be made more economical and sustainable through the valorization and integration of waste product streams obtained at different refining steps.

Keywords: edible oil, chemical, physical refining, valorization, integration

1. Introduction

Refining of crude oil is done to remove unwanted minor components that make oil unappealing to consumers. The conventional methods of oil refining mainly employ chemicals that affect the oil quality, stability, recovery and also generate high volume, low value by-products. After the extraction of oil from oil seeds, refining of this crude oil is performed to remove those impurities that impact the quality of oil. The impurities primarily are phospholipids, free fatty acids (FFA), and minor components like pigments, volatiles, and contaminants. These impurities affect the flavour, color, and stability of the refined oil. The deteriorating effects of these impurities on oil are mentioned in **Table 1** [1].

For crude edible oil, chemical (alkali) and physical refining are the standard processes used in industry. In chemical refining, FFA are removed by neutralization with

Impurities		Deteriorating effects
Phospholipids		Lower oxidative stability
Free fatty acids		Lower oxidative stability, impaired functional properties
Minor Components	Color pigments (chlorophyll and carotenoids)	Lower sensory properties
	Metal salts (iron and copper compounds)	Lower oxidative stability
	Volatile aldehydes and ketones	Off-flavors

Table 1.
Impurities in oil and their deteriorating effects.

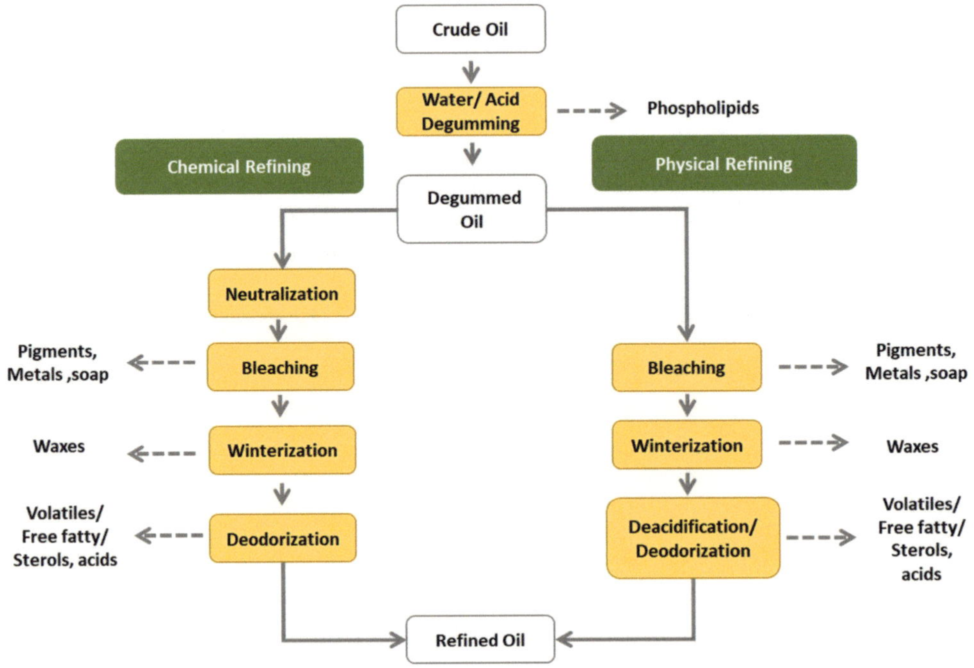

Figure 1.
Basic steps of edible oil refining.

alkali, while in physical refining it is removed by distillation during deodorization. Compared to chemical refining, oil losses are reduced in physical refining. The major steps involved and the components removed in refining steps are shown in **Figure 1**.

2. Chemical refining

Vegetable oil refining consists of multiple steps in sequential order. Based on the oil type, either chemical or physical refining is done.

2.1 Degumming

Degumming is the first step in crude oil refining performed to remove gums. These gums consist mainly of different phosphatides, entrained oil and meal particles (**Figure 2**). The majority of phosphatides present in gums are hydratable. They absorb water and thus become oil-insoluble. Different types of degumming techniques practiced in the refining industry are chemical, enzymatic, and membrane degumming.

2.1.1 Chemical degumming

In chemical degumming, crude oil is treated either with water, acid, ethylene diaminetetraacetic acid (EDTA), etc. Based on the chemicals used, chemical degumming can further be divided into six categories: water degumming, acid degumming, dry degumming, acid refining, organic degumming, and EDTA degumming. Amongst these, water degumming and acid degumming are widely practiced in industry. In water degumming, soft water at the same percentage (2–3%) as the total phospholipids content of crude oil is added to hot oil (70°C) and intensely mixed for 30–60 minutes. It is followed by settling or centrifugation to remove gums and phospholipids. The phosphorus content is typically lowered to 12–170 ppm [2]. In this degumming, only hydratable phospholipids are removed. In industry, acid degumming is done to convert non-hydratable phospholipids (NHP) into phasphatidic acid and calcium/magnesium biphosphate salts. In this process, crude oil is treated with phosphoric acid at a temperature of 90°C [3, 4] to settle the non-hydratable phospholipids. The super acid degumming process of Unilever is carried out at 40°C that produces oil with a phosphorus content lower than that of the standard acid degumming [5]. In the dry degumming process, concentrated acid (70–80%) is added to hot crude oil in an amount of 0.05–1.2%. The acid (phosphoric acid) decomposes the metal salts of acidic phospholipids. In the acid refining process, acid pretreated oil is neutralized

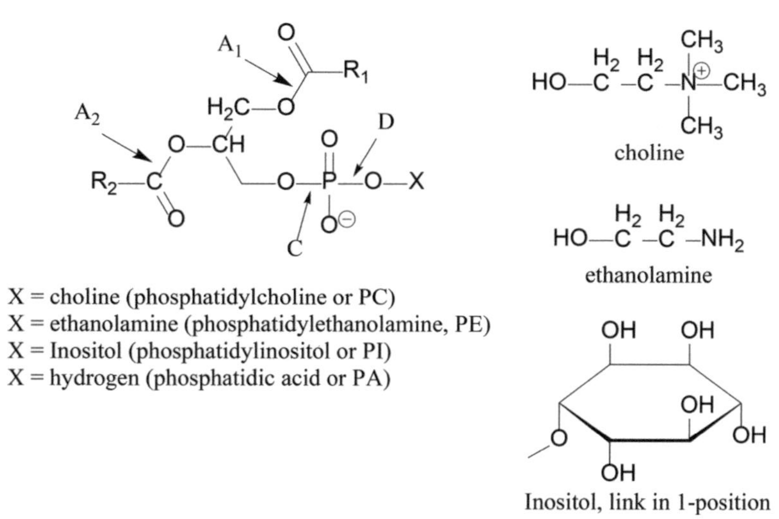

Figure 2.
Different phospholipids (gums).

with caustic. Caustic form soaps have a great emulsifying property that helps in the reduction of NHP. The process helps to achieve a phosphorus level <10 ppm [6].

The process of organic degumming involves the addition of an aqueous solution of organic acid (less than 5% w/w), usually citric acid. It involves a high shear mixing between oil and citric acid solution for less than 30 seconds and a low shear mixing of less than 15 minutes. Thus, three phases are produced; a heavy phase containing citric acid, which can be decanted and recycled; a lighter top phase containing oil; and an intermediate phase containing the gums. This process has the capability to produce oil with a phosphorus content of less than 10 ppm [7]. In the EDTA degumming process, a chelating agent like EDTA in water is added to the oil. In the soft degumming method, a detergent like sodium lauryl sulfate (SLS) is used to assist the contact between the NHP in the oil phase and the chelating agent in the water phase solution. However, there is a complexity in the separation of both phases due to the high stability of the emulsion so formed [8].

2.1.2 Enzymatic degumming

Phospholipids present in crude oil show emulsification properties that affect the oxidative stability of oil. One way to reduce the emulsification properties of phospholipids is to selectively cleave their polar and non-polar parts from one another, and these reactions can be effectively achieved through enzymes. Enzyme catalyzed reactions are selective and occur at moderate temperatures and pH. Phospholipases are the enzymes that are active on phospholipids. Several kinds of phospholipases are reported in the literature and are named based on the position of fatty acid removed from the glycerol backbone, as indicated in **Table 2**. The EnzyMax process was the first enzymatic degumming process adopted in the industry. The process employed the phospholipases A2 sourced from porcine pancreas for the degumming of oil rich in NHP [9]. The reaction was carried out at 50–75°C for 3–4 hours, which resulted in residual phosphorus of 3 ppm in the oil.

PLA1, Lecitase® 10 L and Lecitase® novo was used for degumming of rapeseed oil where phosphorus content was reduced from 280 ppm to 10 ppm [10]. The degumming process developed with enzyme recirculation was successful. The enzyme Lecitase® novo has been used for the degumming of rice bran oil. The phosphorus content and oil losses were high in processed oil due to incomplete reaction [11]. In the enzymatic degumming process, pH is the most critical factor. The pH adjustment of the crude or water degummed oil is done with acid and caustic. The temperature of the reaction mixture is maintained at around 50–60°C and under high shear mixing,

Types of phospholipases	Mode of action
Phospholipase A1 (PLA 1)	Remove the fatty acid from C1- carbon of glycerol backbone
Phospholipase A2 (PLA 2)	Remove the fatty acid from C2- carbon of glycerol backbone
Phospholipase B (PLB)	Acts on lysophospholipid to remove the remaining fatty acid
Phospholipase C (PLC)	Acts on phosphoro-ester group to generated diglycerides

Table 2.
Types of phospholipases.

Enzyme trade name	Enzyme producer	Country of origin	Enzyme activity
Lecitase Ultra	Novozymes	Denmark	Phospholipase A1
Lysomax	Danisco	Denmark	Lipid acyltransferase (Type A2)
Rohalase MPL	AB Enzymes	Germany	Phospholipase A2
GumZyme	DSM	Netherlands	Phospholipase A2
Purifine	DSM-Verenium	Netherlands	Phospholipase C

Table 3.
Commercially available phospholipases for enzymatic degumming.

the enzyme is added, either in pure or dilute form. Enzyme dosing depends on the type of enzyme used and on the phospholipid content of the oil, but usually varies between 50 and 200 ppm. Enzyme recycling is the major constraint that limits its applicant in the commercial refining process. Enzymatic degumming produces side products that affect the oil quality. e.g. Phospholipase A1 liberates a fatty acid from the phospholipid molecule, resulting in a lysophospholipid and a FFA. Commercially available phospholipases for enzymatic degumming are mentioned in **Table 3**.

2.1.3 Membrane degumming

Phospholipids are amphoteric in nature, they tend to form reverse micellar structures in a given medium and possess a molar mass above 20 KDa, with a molecular size ranging from 20 to 200 nm. Ultrafiltration can, thus, be utilized to remove them from oil-hexane mixtures [12, 13]. The degumming of crude sunflower and soybean oils was made possible without the use of solvent, i.e., by using a polymeric ultrafiltration membrane with a 15 KDa molar mass cut-off, 5 bar filtration pressure, 60°C temperature and a filtration flow rate of 0.3 $m^3 h^{-1}$. By this procedure, 77% and 73.5% of phospholipid retention for the sunflower and soybean oils, respectively, was achieved [14]. But, the use of membranes at industrial scale degumming has limitations in membrane stability in organic solvents. Membranes used in organic solvents like hexane generally have a shorter lifespan and cleaning protocols still need to be developed and optimized [13].

2.2 Neutralization

Neutralization step in chemical refining is primarily performed to remove FFA from crude edible oil. Conventionally, caustic is directly added in such an amount that the FFA in crude oil and phosphoric acid used previously during acid degumming are completely neutralized. The alkali reacts with the FFA to form soaps (**Figure 3**) that

Figure 3.
Neutralization reaction.

are further removed by centrifugal separators. But, conventional alkali neutralization has major disadvantages:

- Oil losses due to hydrolysis by alkali and by occlusion in soap stock.
- Soap stock has low commercial value. Splitting of soap stock with concentrated acid generates a heavily polluted stream.
- Water used to wash the oil after alkali neutralization needs to be treated.

In recent advances, nano-neutralization is one of the technologies used for FFA removal. Proven industrial advantages of the nano-neutralization process are significant reductions (up to 90%) in phosphoric and citric acid consumption and a corresponding significant reduction (over 30%) in caustic usage [15].

An adsorption process using an anion exchange adsorbent for the removal of FFA from crude oil is reported [16]. But the adsorbents used for FFA removal have low adsorption capacity per unit of adsorbent, and regeneration is a major concern.

2.3 Bleaching

Bleaching is regarded as a partial or complete removal of color. Bleaching also cleans up the traces of soap, phosphatides, and pro-oxidant metals remaining after caustic neutralization and water washing that hinder filtration, darken the oil, and adversely affect the flavor of the finished oil. Another function, considered primary by many processors, is the removal of peroxides and secondary oxidation products. In conventional processes, adsorbents like neutral earth, activated earth, and activated carbon are used for bleaching. These adsorbents are added in the range of 0.15 to 3% weight of oil. Bleaching is done under vacuum for 15 to 20 minutes at 70 to 110°C [17].

Silica hydrogel has also been employed as an aid to bleaching clay to adsorb soap, residual phosphatides, and trace metals. Silica hydrogel (Sorbil® R927) absorbs soap 1.3 times faster than clay adsorbent. At low residual phosphorus levels in the oil (typically <5 ppm P), the capacity of the silica hydrogel is approximately three times that of the clay. The high adsorption capacity and affinity of silica hydrogels for phospholipids, trace metals, and soaps make these synthetic amorphous silicas ideally suited for use in edible oil refining processes. Silica hydrogel does not adsorb color bodies, such as chlorophyll and carotene, from the oil. Thus, the oil is preferably treated sequentially, first with silica hydrogel and then with bleaching clay. In the first stage, the silica adsorbs the phospholipids, trace metals, and soaps. In the second stage, sufficient bleaching earth is added to remove the color bodies [18]. In addition, use of silica seems to improve bleaching earth filterability, resulting in longer filter cycles and higher presses bleach effect [19].

2.4 Winterization

Some vegetable oils, like sunflower, maize, and rice bran, contain waxes. At low temperatures, these waxes crystallize and result in turbidity in the oil. In the winterization process, the oils are cooled in a simple way, kept at low temperatures of 10-15°C for several hours to crystallize solid-fat fractions. Then the cooled oil passes through a filter to separate waxes and clear oil is obtained [2, 17].

2.5 Deodorization

Deodorization is the last step in edible oil refining. It is basically a vacuum-steam distillation process operated at elevated temperatures to remove FFA and other volatile odoriferous components that cause the undesirable flavors and odors. Additionally, deodorization destroys carotenoid pigments, removes pesticides and cyclopropenoid fatty acids. Deodorization design is influenced by four operating variables, which include vacuum, temperature, stripping rate, and retention time. For a continuous deodorizer, the vacuum, temperature, and retention time lie within the ranges of 2–4 mbar, 200–260°C and 15–150 minutes, respectively [2, 17].

The deodorization step has some negative impact on the nutritional quality of oil. Partial loss of bioactives such as tocopherols, tocotrienols, polyphenols, sterols, and squalene is observed during this step. The high temperature of the deodorization step also results in unwanted side reactions like trans-fat formation, conjugation, and polymerization.

3. Physical refining

Physical refining consists of the same steps described in chemical refining, except for the alkali neutralization process. In this process, alkali is not used for FFA removal; rather, it is removed in the deodorization step by steam distillation. Various physical processes to remove FFAs from oil are reported, like steam refining, inert gas stripping, molecular distillation, hermetic system, and extraction with solvents. On an industrial scale, steam refining is used where superheated steam at 200–270°C is passed over degummed and bleached oil to remove FFA. The advantages of physical refining are that it reduces oil losses and minimizes liquid effluent generation.

4. Value-addition to edible oil refining by-products

Refining produces by-products like gums and deodorized distillate (DOD). Different forms of phospholipids can be obtained from gums. DOD contains compounds like FFA, tocopherols, sterols, and squalene.

4.1 Process for the extraction and purification of phospholipids

The aqueous degumming step during the oil refining process produces wet gums as a by-product. These wet gums, produced during the degumming of crude oils such as soybean, sunflower, canola, maize, etc., are processed to produce lecithin. Chemically, gums are mixtures of phospholipids (±45%), glycolipids (±10%), glycerides (±40%) and sugars (±10%). Soybean oil is the primary commercial source of vegetable lecithin. The main phospholipids (PL) present in soya are phosphatidylcholine (PC), phosphatidylethanolamine (PE), phosphatidylinositol (PI) and phosphatidic acid (PA) [20]. PL are polar lipids with hydrophilic and hydrophobic parts and have wide applications in food as emulsifiers, viscosity regulators, anti-spattering and dispersing agents. These PL are also used as nutraceuticals and/or added to animal feed for nutritional enhancement [21, 22].

Different processes have been utilized for isolation, purification or enrichment of PC from natural lecithin, such as: liquid–liquid extraction, supercritical fluid extraction, solvent fractionation, and adsorption. Based on the polarity of the head group, a solvent like ethanol is used for the fractionation of different classes of phospholipids. The PC fraction of lecithin is enriched by ethanol extraction and is very useful as an emulsifier of oil in water emulsions [23]. Lecithins modified by enzymatic hydrolysis, acetylation, and alcohol fractionation processes give a range of food grade emulsifiers with different hydrophilic–lipophilic-balance (HLB) values [24].

4.2 Purification of tocopherols (TC) and Tocotrienols

Tocopherols and tocotrienols are natural forms of vitamin E, each exhibiting four different isoforms as α, β, γ and δ. Vitamin E is an essential fat-soluble vitamin with antioxidant activity. There is growing interest in natural forms of vitamin E because they are promising compounds for maintaining a healthy cardiovascular system and blood cholesterol level [25]. α-TC is the predominant form of vitamin E in tissues, and low intake results in vitamin E deficiency-associated ataxia. Recent mechanistic studies combined with preclinical animal models indicate the tocotrienol subfamily possesses powerful neuroprotective, anticancer, and cholesterol-lowering properties [26]. Tocopherols are present in major vegetable oils, while tocotrienols are present in palm oil, rice bran oil, and annatto seed. For commercial production of natural tocopherols and tocotrienols, enrichment and purification of TC and TT from deodorizer distillate (DOD) and fractionation of palm oil are practiced in industry.

In the refining process of edible oil, deodorization is the final step carried out at high-temperature, high-vacuum steam-distillation to produce high-quality refined oil. The deodorization process removes FFAs and volatile aldehyde and ketone compounds responsible for the off-flavor and odor of oil. Along with impurities, the deodorization process also partially removes tocopherols, tocotrienols, and phytosterols. So, DOD is a complex mixture of tocopherols, tocotrienols, phytosterols, glycerides, hydrocarbons, and FFA [27]. Tocopherols and tocotrienols have been purified from DOD by a combination of molecular distillation, ethanol fractionation, chemical alcoholysis, and ion exchange chromatography [28]. Preparation of high purity concentrates of TC and TT involves a series of physical and chemical treatment steps like hydrolysis, neutralization [29], supercritical extraction with multistage counter–current column [30], trans-esterification to form methyl esters followed by fractional distillation, membrane separation, enzymatic esterification [31] and batch adsorption and desorption using silica [32].

5. Conclusion

Refining of edible oil by chemical or physical processes aims towards removing the undesirable impurities that affect oil quality and stability. Understanding the chemical nature of oil and associated impurities is important to designing refining steps. The recent refining advance has resulted in minimizing chemical usage, oil losses, and effluent generation. With the rising demand for natural emulsifiers and antioxidants, the by-product streams of gums and DOD can be utilized to obtain high-value products like phospholipids and tocopherols, respectively. There is a need for technological advances that integrate refining and valorization of refining by-products, making the edible oil industry more economical and sustainable.

Author details

Anup Sonawane[1*] and Samadhan R. Waghmode[2*]

1 DBT-ICT Centre for Energy Biosciences, Institute of Chemical Technology, India

2 Department of Microbiology, Elphinstone College, Mumbai, India

*Address all correspondence to: anoopsonawane@gmail.com
and samadhanwaghmode@gmail.com

References

[1] Čmolík J, Pokorný J. Physical refining of edible oils. European Journal of Lipid Science and Technology. 2000;**102**(7):472-486

[2] Bailey AE, Shahidi F. Bailey's Industrial Oil and Fats Products. Canada: John Wiley and Sons; 2005

[3] Mag TK, Reid MP inventors; Canada Packers Inc, assignee. Continuous process for contacting of triglyceride oils with_an acid. United States patent US 4,240,972. 1980

[4] Campbell SJ, Nakayama N, Unger EH. Chemical degumming of crude vegetable oils. United Oilseed Products Ltd, Canadian Patent. 1983;**1**(157):883

[5] Ringers HJ, Segers JC, inventors; Lever Brothers Co, assignee. Degumming process for triglyceride oils. United States patent US 4,049,686. 1977

[6] Dijkstra AJ. Enzymatic degumming. European Journal of Lipid Science and Technology. 2010;**112**(11):1178-1189

[7] Copeland R, Belcher WM. Improved method for refining vegetable oil. PCT Patent Application WO 00/31219 assigned to AG Processing Inc. 2000

[8] Jamil S, Dufour JP, Deffense EM, inventors; Fractionnement Tirtiaux SA, assignee. Process for degumming a fatty substance and fatty substance thus obtained. United States patent US 6,015,915. 2000

[9] Aalrust E, Beyer W, Ottofrickenstein H, Penk G, Plainer H, Reiner R, inventors; Metallgesellschaft AG, Evonik Roehm GmbH, assignee. Enzymatic treatment of edible oils. United States patent US 5,264,367. 1993 Nov 23

[10] Munch EW. Degumming of Plants Oils for Different Applications. Cairo: Society of Chemical Industry; 2007

[11] Ferreira ML, Tonetto GM. Enzymatic Synthesis of Structured Triglycerides: From Laboratory to Industry. Argentina: Springer; 2017. p. 35-54

[12] Koseoglu S. Advantages of membrane degumming—Real or imagined? European Journal of Lipid Science and Technology. 2002;**104**(6):317-318

[13] Cheryan M. Membrane technology in the vegetable oil industry. Membrane Technology. 2005;**2**:5-7

[14] Koris A, Vatai G. Dry degumming of vegetable oils by membrane filtration. Desalination. 2002;**148**(1-3):149-153

[15] De Greyt WF. Current and future technologies for the sustainable and cost-efficient production of high quality food oils. European Journal of Lipid Science and Technology. 2012;**114**(10):1126-1139

[16] Sari A, Iþýldak Ö. Adsorption properties of stearic acid onto untreated kaolinite. Bulletin of the Chemical Society of Ethiopia. 2006;**20**(2):259-267

[17] Hamm W, Richard JH, Gijs C, editors. Edible oil processing. Hoboken, NJ, USA: Wiley-Blackwell; 2013

[18] Nock A. Silica Hydrogel and its Use in Edible Oil Processing. USA: The American Oil Chemists' Society; 2010

[19] Jalalpoor M. PCT patent application 2008/02552 A2, 2008

[20] Scholfield CR. Composition of soybean lecithin. Journal of the American Oil Chemists' Society. 1981;**58**(10):889-892

[21] Ceci LN, Constenla DT, Crapiste GH. Oil recovery and lecithin production using water degumming sludge of crude soybean oils. Journal of the Science of Food and Agriculture. 2008;**88**(14):2460-2466

[22] Van Nieuwenhuyzen W. The industrial uses of special lecithins: A review. Journal of the American Oil Chemists Society. 1981;**58**(10):886-888

[23] Holló J, Perédi J, Ruzics A, Jeránek M, Erdélyi A. Sunflower lecithin and possibilities for utilization. Journal of the American Oil Chemists' Society. 1993;**70**(10):997-1001

[24] Van Nieuwenhuyzen W. The changing world of lecithins. Inform. 2014;**25**(4):254-259

[25] Colombo ML. An update on vitamin E, tocopherol and tocotrienol—Perspectives. Molecules. 2010;**15**(4):2103-2113

[26] Sen CK, Khanna S, Rink C, Roy S. Tocotrienols: The emerging face of natural vitamin E. Vitamins and Hormones. 2007;**76**:203-261

[27] Shimada Y, Nakai S, Suenaga M, Sugihara A, Kitano M, Tominaga Y. Facile purification of tocopherols from soybean oil deodorizer distillate in high yield using lipase. Journal of the American Oil Chemists' Society. 2000;**77**(10):1009-1013

[28] Ghosh S, Bhattacharyya DK. Isolation of tocopherol and sterol concentrate from sunflower oil deodorizer distillate. Journal of the American Oil Chemists' Society. 1996;**73**(10):1271-1274

[29] Chu BS, Baharin BS, Quek SY, Man YC. Separation of tocopherols and tocotrienols from palm fatty acid distillate using hydrolysis-neutralization-adsorption chromatography method. Journal of Food Lipids. 2003;**10**(2):141-152

[30] Brunner G, Malchow T, Stürken K, Gottschau T. Separation of tocopherols from deodorizer condensates by counter current extraction with carbon dioxide. The Journal of Supercritical Fluids. 1991;**4**(1):72-80

[31] Ramamurthi S, Bhirud PR, McCurdy AR. Enzymatic methylation of canola oil deodorizer distillate. Journal of the American Oil Chemists' Society. 1991;**68**(12):970-975

[32] Chu BS, Baharin BS, Man YC, Quek SY. Separation of vitamin E from palm fatty acid distillate using silica: I equilibrium of batch adsorption. Journal of Food Engineering. 2004;**62**(1):97-103